GERARD C. SMITH

A Career in Progress

GERARD C. SMITH

A Career in Progress

Edited by

Henry Owen

and

John Thomas Smith II

University Press of America,® Inc.

4720 Boston Way
Lanham, MD 20706

3 Henrietta Street
London WC2E 8LU England

Printed in the United States of America

British Cataloging in Publication Information Available

Library of Congress Cataloging-in-Publication Data

Gerard C. Smith : a career in progress.

1. Nuclear arms control—United States. 2. Smith, Gerard C. I. Owen, Henry, 1920– .
II. Smith, John Thomas, 1943– .
JX1974.7.G44 1989 327.1'74'0973 89-5774

ISBN 0–8191–7444–0 (alk. paper)

All University Press of America books are produced on acid-free paper.
The paper used in this publication meets the minimum requirements of American National Standard for Information Sciences—Permanence of Paper for Printed Library Materials, ANSI Z39.48–1984. ♾

Contents

Preface

Last summer Henry Owen suggested that we collaborate in the organization of a *Festschrift* in honor of Dad's 75th birthday. The dictionary defines *Festschrift* as "a volume of writings by different authors presented as a tribute . . . especially to a scholar." The word is a combination of two German words "fest" meaning festival or celebration and "schrift" meaning writing. While a *Festschrift* can consist of formal essays on an academic topic, Henry's idea was that this volume comprise informal, personal reflections upon my father's ongoing public service career.

I enthusiastically endorsed Henry's suggestion and have enjoyed the privilege of collaborating with him in its execution. I have not acted upon Henry's urging that I produce a chapter regarding Dad's role as *pater familias.* In the first place, I didn't dare risk the criticism of my three siblings, Sheila, Gerard and Hugh, each of whom has her or his own warm perceptions of Dad, which I could not presume to capture. In the second place, Dad is a sufficiently self-effacing and private man that he would be embarrassed by extensive public discussion of private matters. Nevertheless, I venture a few observations on this score—without serious risk of dissent by my siblings and, I hope, without embarrassment to Dad.

From the time each of us reached the age of reason, to the present day, Dad has led us by example rather than edict. As a result, none of us has rebelled, and each has tried—with imperfect success—to emulate his powerful example.

Each of us tries to be a loyal, caring and considerate spouse and parent. We care deeply about public and international affairs and are trying to impart this example to our own children. While none of us is as deeply, thoughtfully and consistently religious as Dad, his example helps keep us in touch with the Church. In our family, professional, and social dealings, we try to display at least a fraction of the courtesy, patience and generosity that Dad consistently displays.

Because of the force of his example and his involvement in matters of high national importance, Dad was a somewhat imposing and austere figure to each of us when we were too young to appreciate his unflagging humor and *joie de vivre*. As we matured, we have come to enjoy Dad as a friend and even a "best friend," while he has nonetheless retained a moral authority never held by a "mere" friend.

It is interesting that many of the attributes that make Dad a special father are ones that are noted in this book by his friends and colleagues as they discuss their associations with him. This coincidence is clear testament to the continuity and integrity between Dad's public career and private life. Indeed, a number of the authors have not simply been professional colleagues of Dad's, but are also friends of the family known to and deeply admired by all of us.

Robert Bowie preceded Dad as head of Policy Planning at the State Department. Like Dad, he has dedicated his career to public service and education regarding vital international issues. He is a close family friend and Eastern Shore neighbor. There is no one better suited to compose a foreword to this volume.

John Avirett, or "Uncle John" as we "children" know him, has been Dad's friend for more than 45 years. After naval service, he practiced law in Baltimore and became the managing partner of that city's pre-eminent firm, Piper & Marbury. We owe him an incalculable debt for his role in steering Dad to the acquisition of Ratcliffe Manor on Maryland's Eastern Shore which, nearly 45 years later, remains the family's geographic center.

William McGuirk labored side-by-side with Dad in the halcyon days of the Atomic Energy Commission. While we have seen him too seldom since the early 1950s, he has remained a family exemplar. A man of unbounded energy, he has, among other things, served as head of the Chemical Division of W.R. Grace & Company, President of the Mercantile Bank in Baltimore, and publisher of the *Baltimore Sun*. He has also made time to raise 15 children.

Henry Owen has been one of Dad's closest friends and confidants since their joint service on the State Department's Policy Planning Staff from 1957–61. Henry, himself, became chairman of this important staff, succeeding Walt Rostow in 1965. Subsequently, Henry was a fellow at the Brookings Institution, contributed insightful commentary to the *Washington Post's* Op-Ed page, and served as an Ambassador-at-Large during the Carter administration. Dad and Henry are partners in the consulting firm that they jointly founded. Through an association with Salomon Brothers, Henry collaborates in matters of high international finance with my brother Gerry. He is an affectionate friend to all of us.

Admiral John Marshall ("Squidge") Lee worked with Dad on the ill-fated multilateral force during the Kennedy and Johnson administrations, and returned for more punishment, serving under Dad as an Assistant Director of the Arms Control and Disarmament Agency (ACDA). He remains an active and vigorous conscience for rational arms policies. Dad has introduced his children to a number of admirals, including the formidable Rickover. None can match "Squidge" Lee for zest and erudition.

Lawrence Weiler served as "counselor" to Dad when he was the Director of the Arms Control and Disarmament Agency. His contribution to this volume, while generous to Dad, also illuminates his vital assistance to Dad at ACDA, as an institutional memory and as conscience and gadfly.

Harold Brown has generously contributed his recollection of the SALT I negotiations. This prodigious scholar and former Secretary of Defense continues public instruction on the need for rational arms policies. He has also brought unmatched intellect to games of charades on New Year's Eve in the Smith household.

Max Kohnstamm, presently the Secretary General of the Action Committee for Europe, has long shared Dad's interest in the strengthened cooperation among Europeans and in the partnership between such a strengthened Europe and the United States. He is one of a number of influential Europeans who have been friends of Dad's and acquaintances of the family. He and his wife joined the Smith family in a cruise of the Yugoslav coast in the summer of 1968.

Phil Farley with typical diligence and lucidity has contributed a cogent essay regarding Dad's role in the nation's efforts at nuclear nonproliferation. This self-effacing and incisive public servant is admired by all of us. We saw the most of him when he ably assisted Dad as his deputy at ACDA. We miss him since his removal to Stanford University.

Spurgeon Keeny, like Dad, has long labored for the cause of rational nuclear policies and served as Assistant and later Deputy Director of ACDA. He is a holder of the Rockefeller Award for Distinguished Public Service. As President of the Arms Control Association, Spurgeon continues his close collaboration with Dad. We Smiths have known and admired Spurgeon for years.

Father Paul Mohan is a professor at the Catholic University of America. He contributes an appreciation of Dad's life-long commitment to the Catholic faith and the institution of the Church. Through our family's Catholicism, we have met a number of highly intelligent and articulate priests—but none more so than Father Mohan.

Two other friends and colleagues of Dad's have contributed short reflections on their associations with him. Their pieces are included as an appendix.

Bob Schaetzel, a distinguished public servant and former U.S. Ambassador to the European Economic

Community, began assisting Dad with problems of harnessing the atom in the 1950s. Their collaboration continues to this day. I have "enjoyed" a drubbing administered by this keen and skillful tennis player.

James Chace acted as Editor of *Interplay* magazine, which Dad founded and nurtured during the late-1960s. A distinguished author and scholar, Mr. Chace has subsequently served as Managing Editor of *Foreign Affairs*. Like many of Dad's colleagues, he radiates a deep and intellectual concern for an improved international polity.

The authors have given generously of their time to make this volume possible. Their own distinction as well as their friendship compliment Dad beyond any of the writings herein.

John Thomas Smith II
Washington, D.C.
February 1989

Foreword

Jean Monnet, who devoted twenty-five years to the European Community, once said to George Ball: "George, you should stop diffusing your energies. You should select a single, great objective and concentrate on it until it is accomplished. You may have to make short-term tactical detours, but you must never lose sight of your central goal, even when the road ahead seems hopelessly blocked."

Gerry Smith's career has exemplified Monnet's principle. Since 1950, when he started work with the Atomic Energy Commission, he has dedicated his talents to the "single great objective" of controlling nuclear weapons. Observing nuclear tests had convinced him that this was "the most important business in the world."

For nearly forty years he has been concerned with the "long Soviet-American postwar process of trying to adjust to the fact . . . that each has the power to destroy the other and in the process to destroy world civilization." Neither detours nor road-blocks have diverted his sight or energy from his central goal.

In its pursuit he served in a wide variety of posts under five Presidents. His four years with the AEC were the ideal apprenticeship. They came at the opening of the nuclear era as we have known it. Only then did the U.S. begin to accumulate a significant stockpile of nuclear weapons. The first Soviet test had occurred in late 1949. And Truman had then greatly expanded U.S. facilities and authorized the hydrogen bomb program. Thus, as personal assistant to Commissioner Murray, Gerry directly participated in the analysis and decision-making of the AEC during this formative period.

In the State Department in the Eisenhower administration, first as Special Advisor to Secretary Dulles on nuclear matters, and then as my successor as Director of Policy Planning, he took part in confronting the new choices and dilemmas posed for foreign policy by the atom. Eisenhower's own deep interest in finding ways to mitigate or control the risks of nuclear war created a stimulating context for exploring a variety of means to that end: Atoms for Peace, arms control of various kinds, safeguards against surprise attack, the International Atomic Energy Agency, and the Multilateral Force for NATO to discourage national nuclear forces. The results of these efforts were meager; conditions were not propitious. But this decade of experience equipped him well for leading the Arms Control and Disarmament Agency and the SALT delegation under Nixon, and for his work as Ambassador-at-Large for Non-Proliferation under Carter.

His varied activities have thus involved him in every aspect of the nuclear problem: the role of nuclear weapons in military strategy (flexible response, MIRVs, strategic defense); constraining the spread of nuclear weapons (IAEC, the NPT, Carter's Special Advisor); arms control (ACDA, SALT I, ABM); and education on all these issues (Arms Control Association).

Out of office, Gerry has been unflagging in his efforts to promote arms control by speeches, testimony and articles, and by heading the Arms Control Association. When the Reagan administration sought to undermine the ABM treaty in the pursuit of the Strategic Defense Initiative, Gerry assisted through such means the successful campaign led by Senator Sam Nunn to block that cynical wrecking action.

Gerry's career has spanned almost the entire period of the cold war, from its post-war beginnings to what may well become its terminal phase. In that time, arms control has evolved from a peripheral aspect of foreign policy into a central factor in the effort to move toward more normal East-West relations. Gorbachev seems to be opening the way toward much lower armaments, nuclear and conventional, and more stable and secure peace. Whether a basic change in relations can be achieved over the next few years remains to be seen. But the U.S. and its allies will be derelict in their obligations to their people if they fail to explore and test these opportunities actively and constructively. The progress thus far is, of course, the result of historic forces and the efforts of countless individuals, but Gerry's activities have clearly made their contribution.

As the essays in this volume testify, Gerry's strength has been his character and dedication. He has worked assiduously to master the specifics of these complex issues. While drawing on the knowledge and advice of experts, he reaches his own conclusions, based on his own good judgment. He is committed to his goals, but he relies on reason and facts, not zealotry. Under Carter, while the statute to constrain proliferation depended heavily on coercion and penalties, he recognized that positive results would require cooperation, which would be best achieved by trying to remove or assuage some of the concerns prompting nations to seek nuclear weapons; and he sought to adapt the program accordingly, as far as feasible. He can be tenacious and firm, as well as courteous and patient.

His greatest asset is integrity—in the literal sense. He is all of a piece, without pretense. Hence he is at ease with himself. His inner serenity is surely rooted in his quiet religious faith and was long bolstered by the warm support of his wife, Bernice. His solid honesty generates respect and trust from those with whom he negotiates or deals. And in colleagues and collaborators, it has inspired the deep loyalty and affection expressed throughout these essays.

In each stage of our history, the nation has been served by civic-minded citizens who have been motivated not by the desire for public acclaim or profit but by the satisfaction of having contributed to its welfare or safety. Gerry is a worthy exemplar of that tradition.

Robert R. Bowie
Washington, D.C.
February 1989

Introduction to Washington

Ensign Gerard C. Smith, USNR, reported for duty in Washington in 1941 and was assigned to service expediting procurement of essential war materiel. He served in that capacity with increasing responsibilities until the end of the war when he was released to inactive duty as a lieutenant, USNR.

The atmosphere in Washington during those days was indeed strange. A sort of businesslike quiet prevailed. Nothing exciting—just a recognition that there was important work at hand, and a determination to get on with it, to get it done and get back home. To do the job, thousands of highly competent lawyers, businessmen and other professionals had been recruited from all over the country, on leave from their various firms and companies. Some old friends from law and business schools met again, and new friends were made. Most of them were in uniform, and one of their problems was learning to understand and work with the Regulars under their system in which rank was important. As the saying goes, "R.H.I.P.—Rank has its Privilege." On the other hand, the Reserves were for the most part under-ranked in relation to their competence in their respective fields and were very much out-ranked by their opposites in the Pentagon.

It was into this rather grim atmosphere that Gerry came as an ensign, USNR. He had graduated from Yale Law School, with the intention of practicing law in New York with his father, then Vice President and General Counsel for General Motors. The law firm name was to be a quaint old fashioned "John Thomas Smith and Son". The intention to establish the firm was delayed by the war and, due to the untimely death of Mr. Smith, was never fully realized.

In early 1943 I reported for duty in Washington and was assigned to service in the Navy's Office of General Counsel, which developed into a top drawer law firm established at the insistence of Secretary Forrestal to handle all legal aspects of the vast procurement program for which he was responsible. In the course of this duty, I met Gerry for the first time and, despite the substantial difference in age, we soon became good friends.

In the course of Gerry's wartime work in Washington, many things happened which influenced or came to be reflected in his post-war life and career. Two of them were memorable. One dealt with his purchase of "Ratcliffe Manor" near Easton, Maryland; the other with a paternal prophecy made to him by his father when he was living in Southampton, New York.

As for "Ratcliffe Manor", Gerry, who was always an avid sailor, had heard much about the Chesapeake Bay country and, being so close to it, was anxious to see some waterfront property. On one occasion in 1945, he asked me if I knew of a place called "Ratcliffe Manor" which was for sale. I told him that indeed I did; that it was a lovely estate and that if he was interested, he really should go and see it. So, on one of our few free days, we drove over to the Eastern Shore and saw the property with its beautiful Georgian manor house and magnificent boxwood garden located on approximately 500 acres of fertile fields and bordering on the shores of the peaceful Tred Avon River with a boat dock not more than an "eight iron" shot from the house. The real estate agent showed Gerry through the house. When they came out, I could see that Gerry was worried. He hastened to tell me that he liked the place so much he wanted to buy it; but hesitated to do so until his wife Bernice could see it and approve. The other hitch was that the agent was on the verge of closing a deal with a very interested prospect who was to see it again that very afternoon. Suppose he bought the estate and Bernice did not approve? What should he do?

By that time I had come to know Gerry well enough to be quite straightforward, and I said, "Gerry, I'll be frank. If you buy "Ratcliffe Manor" and Bernice doesn't like it, you can well afford to sell it, even at a loss. If you don't buy it and Bernice likes it as much as I am sure she will, you'll really be in the 'dog house'." So he signed on the dotted line, Bernice more than approved, and "Ratcliffe Manor" became the family estate of the Gerard Smith clan. Here the family rallies from time to time, and it is here that Gerry can and does get away from it all, to read, reflect and relax; and to entertain his many State Department, government and business friends and associates including his many distinguished guests from abroad.

As for the "paternal prophecy," Gerry and I frequently went to New York on navy business, and on one such occasion when we had finished the job at hand we went up to Southampton to see his father who at that time was living there with his daughter, Maureen Shanley. On arrival, we hurried to his father's study, a large unique circular room at the very top of a country farm silo. We climbed the steps and there in his remarkable study was Mr. Smith reading a book. After I had been introduced, Gerry asked his father what was the book in which he was so engrossed. To our surprise, Mr. Smith, Vice President and General Counsel of General Motors, said, "Karl Marx, *Das Kapital*" and added, "Young men, you and all your generation should read and digest this book because you will certainly be 'dealing with the Russians' ". Little did he realize the extent to which Gerry would in due course be "dealing with the Russians."

In retrospect, Gerry's wartime experience in Washington, working in and with the military establishment and government officials in other departments gave him an insight into how things were accomplished in the Capital, caught his interest and helped prepare him for what has been a brilliant career of 20 years important government service under five post-war presidents. It has included various major State Department posts, culminating in his becoming Director of the United States Arms Control and Disarmament Agency, an Ambassador-at-Large and Special Representative of the President and as Chief U.S. Delegate to the SALT I negotiations with the Russians, resulting in the ABM Treaty which was not only signed but also approved by the Senate—a real breakthrough for better relations between the United States and Russia. How prophetic Mr. Smith had been when he told Gerard that he would be "dealing with the Russians".

The "Ratcliffe Manor" and the "paternal prophecy" episodes were but two of many I was privileged to share with Gerry as a result of our being brought together in our wartime service in Washington. As we saw more and more of each other, there developed a close and personal friendship with Gerry and his family—a friendship which my wife Barbara and I will always cherish and for which we shall be forever grateful.

John Avirett, II
Baltimore, Md.
October 1988

Introduction to Nuclear Matters: *The Atomic Energy Commission, 1950–1954*

On March 20, 1950, from the vacation White House in Key West, President Truman announced the appointment of fifty-nine-year-old Thomas E. Murray to fill the vacancy on the Atomic Energy Commission created by the resignation of the chairman, David Lilienthal. Murray's curriculum vitae in *Who's Who* showed him to be an engineering graduate of Yale University, president of a substantial family-owned company making electrical protective devices and automobile parts, and a director of a number of companies, including Chrysler Corporation and The Bank of New York. A holder of more than 200 patents, Murray was an intensely creative, if unorthodox thinker, highly intuitive, quick to reach a decision and impatient of bureaucratic delay. Murray also possessed the asset so greatly prized in Washington circles, ready access to the president—an access that from time to time saw quarter-hour appointments transmuted to two hours of friendly chit-chat—to the consternation of the presidential staff. Murray was quickly confirmed by the Senate. Upon discovering the awesome mixture of minutiae and matters of monumental import coming before the commission in almost daily meetings, Murray quickly turned to Gerard Smith and asked him to serve as his executive assistant. Thus was born a partnership "made in Heaven"—a partnership which diverted Gerard from corporate law to a career in government service and may have cost the Supreme Court a distinguished justice or General Motors the leadership which might have saved it from Japan. Undaunted by long days in the Commission that ordinarily ran from seven 'til seven, Gerard set that jutting jaw in indication of a fierce determination to master every aspect of the atomic world and to buttress Tom Murray's intuitive brilliance with detailed knowledge, studied perceptiveness and a remarkable ear for the ring of truth. Almost fifty years ago, in a vastly different context,* Gerard quoted the Latin exhortation: "*ingredi totus, maneri solus, egredi alius.*" He went into the AEC with his whole heart and mind, he worked without benefit of any support save his own intellectual ability and emerged four years later with the depth of knowledge of the world of the atom which was to serve him and us so well in the years to come.

Memories being what they are, it might be well to briefly set the scene in the nuclear world in the spring of 1950 when the Murray-Smith team came to the AEC. Bernard Baruch's efforts at international control of the atom having fallen to the Russian Security Council veto,

*All At Sea by Gerard C. Smith, 1939

the congress passed the Atomic Energy Act of 1946 creating a five-man commission to assume responsibility for the facilities of the Manhattan Engineering District and to develop the peaceful uses of the atom. From the Manhattan District, the Commission fell heir to offices, laboratories and nuclear production plants in 13 states, 2,000 military personnel, 4,000 civilian government employees and 33,000 contractor employees. The stockpile of nuclear weapons inherited by the Commission was illusory, forcing the Commission to report to the president on April 3, 1947, three months after assuming its responsibilities, that in fact there were no assembled weapons available—the nuclear deterrent on which the West rested its hopes was virtually non-existent.

From this nadir of nuclear preparedness, the Commission slowly emerged during the next two and half years. Progress was inhibited by indecision stemming, in part, from conflicting advice from its statutory advisory bodies, the Military Liaison Committee and the General Advisory Committee which was made up of eminent scientists and engineers. In addition, the Joint Congressional Committee on Atomic Energy, under the leadership of Senator Hickenlooper, did little to support the Commission in its battle for appropriations and in its repeated conflicts with the House Un-American Activities Committee under the chairmanship of J. Parnell Thomas.

A new era in the Commission's history began on September 23, 1949, when President Truman announced that the U.S. Long Range Detection System had picked up irrefutable evidence of a Soviet nuclear explosion and pointed out, "the necessity for that truly effective enforceable control of atomic energy which the government and the large majority of the United Nations support."

On February 2, 1950, the commissioners were informed that German born British scientist, Klaus Fuchs, had confessed to espionage. Fuchs had been a member of the British team working on fission weapons at Los Alamos during World War II and had participated in discussions there about the theory of thermonuclear "super" weapons.

Immediately, the Commission, the Department of Defense and the Joint Committee began separate critical re-examinations of the rate and scale of our nuclear program. Guided by these studies which were triggered by the Russian explosion and Fuchs' defection, the president ordered sharp increases in the production of fissionable material, a major effort to develop a thermonuclear weapon and highest priorities for military propulsion reactors for submarines and aircraft.

During the regime of its first chairman, David Lilienthal, the Commission had delegated much of the major decisionmaking to the general manager, Carroll Wilson. Wilson, in turn, had worked closely with the Military Liaison Committee and the General Advisory Committee in making both operational and policy decisions. The role of the Commission was somewhat analogous to that of a corporate board of directors—selecting management and setting broad policy. Unlike a corporate board, the Commission had little input in establishing the goals and objectives of the nuclear program.

---

*Public Papers of Harry S. Truman, 1949 p. 485

Following Lilienthal's resignation and a brief interregnum in which Senator Pike acted as chairman, on July 11, 1950, Gordon Dean, former law partner of Senator Brien McMahon, chairman of the Joint Congressional Committee, was named as chairman of the Atomic Energy Commission by President Truman.

Dean was dedicated to putting the Commission into an active managerial role and McMahon, ably assisted by his executive director, William L. Borden, was equally dedicated to making the Joint Committee an informed and intensely active oversight body.

The outbreak of hostilities in Korea on July 5, 1950, reinforced President Truman's decision to sharply expand the nuclear effort. The Murray-Smith team was faced with the necessity of playing a part in managing a wartime effort to rebuild our nuclear deterrent, to develop a "super" thermonuclear weapon, and to design and construct reactor power plants suitable for submarine propulsion.

Gerard was frequently asked what he did as executive assistant to Tom Murray. His stock answer (typically self-effacing) was "hew wood and haul water." Were his description accurate, he would have been excoriated as a denuder of forests and a despoiler of wet-lands. At this expansive stage in the Atomic Energy Program, there were about 300 voting agenda decisions made by the Commission per year. These agenda decisions involved a spectrum that varied from authorizing capital expenditures totaling several billion dollars to denying security clearance to Robert Oppenheimer, a nationally known nuclear physicist. During twelve-hour days, Gerard reviewed the entire written record on each of these agenda items, personally conferring with the responsible senior members of the Commission staff, and frequently conferring with associates in State and Defense when those departments' interests were involved in AEC matters. Then, prior to each Commission meeting, Gerard briefed Tom Murray, whose alter ego he was, suggesting a course of action to be pursued. This was a Herculean labor in itself.

Gerard also took upon himself the responsibility for probing deeply into particular areas in which he felt the Commission's efforts were inadequate. For example, Gerard perceived that the entire effort to increase the rate and scale of the National Nuclear Energy Program, as directed by the president, was dependent upon a rapidly depleting supply of uranium ore from the Belgian Congo. Gerard recognized the necessity for striving on a practical basis to obtain ore from any available source. Working through Tom Murray, he drove to bring South African mines into production of uranium ore from gold mine tailings. In November 1950, the combined development agency (U.S., U.K. and Canada) concluded an ore purchase agreement with South Africa that was in no small measure a product of Gerard's research and initiative.

Appreciating the premium value of a domestic supply of ore, Gerard worked unceasingly with Jesse Johnson, Director of the Raw Materials Division, to establish incentives for exploration and development in the American West. The conventional wisdom was that the U.S. had very favorable pricing in its uranium purchase contract with the Belgian Congo and, therefore, no higher prices should be paid for domestic ore until the inexpensive Congo supply ran down. With hundreds of millions of dollars going into expansion of production plants wholly dependent upon ore supplies, it is difficult to believe that such short term "wisdom" could have prevailed. To the delight of Jesse Johnson and the mining fraternity of the southwest, Gerard destroyed the arguments of the "stand patters." In time, domestic production exceeded all expectations.

In researching world production of uranium ore, Gerard unearthed a report on a modest deposit in Spain. Wishing to leave no uranium source, however remote, unexplored, he suggested to Mr. Murray a visit to General Franco as the only practical way to quickly determine the possibilities of bringing the Spanish deposit into production. The appropriate "through channels" request was made to the State Department and our ambassador in Madrid was turned down summarily by the Spanish authorities. Gerard proposed that a call by Mr. Murray to the appropriate party in the Vatican might produce a different result. Forty-eight hours later, our ambassador was summoned and told that the Generalissimo would like to meet an American atomic energy commissioner named Murray at his earliest convenience.

Norman Polmar and Thomas B. Allen in their book *Rickover—Controversy and Genius* speak of Murray as "the commissioner with whom Rickover developed the closest rapport."* Gerard himself states that before accepting President Nixon's appointment as director of the Arms Control and Disarmament Agency, Admiral Rickover was one of three friends whose advice he sought. Indeed,

---

*Page 151

there was a strong bond of mutual respect between Gerard and "Rick." Captain Rickover had not been selected for promotion to rear admiral by the 1951 or 1952 selection boards and was therefore destined for retirement. Gerard learned of this from Commander James Dunford, USN, who was a part of Rickover's original nuclear team of four naval officers and was later to serve as an assistant to Mr. Murray. Recognizing Rickover's unique ability to expedite the Submarine Reactor Program, Gerard suggested that Mr. Murray alert President Truman to the fact that Rickover would soon be lost to the program and that such a loss would be disastrous. Mr. Murray so advised the president in one of his fifteen-minute appointments that lasted two hours. it was the opening salvo in a battle that, in conclusion, saw Rickover on the cover of *Time* as a rear admiral and the Navy brass retreating in disarray.

Let there be no mistake about the fact that Gerard, working intimately with Commissioner Murray, had a powerful impact on the entire program of the AEC which, in accordance with President Truman's directive, was heavily oriented toward military applications.

As time wore on, because of the wartime military emphasis of the Commission's program, Gerard's duties exposed him increasingly to the thinking of the Military Liaison Committee and the strategic planners in the Department of Defense. As the stockpile of "conventional" fission weapons grew and as the "super" thermonuclear weapon progressed from theory to "laboratory demonstration" to a weapon in the arsenal of democracy, Gerard's common sense led him to ask repeatedly "how much destructive power is enough?" As his concern grew, he refused to become calloused to a U.S. defensive strategy that called for the widespread use of strategic nuclear weapons against an aggressor. In 1951, he journeyed to Eniwetok for the greenhouse nuclear tests. Gerard observed that witnessing a test was like "having a look at Hell." This voyage to the Pacific brought to full maturity Gerard's conviction that nothing in the modern world approached in importance the need to control nuclear weapons and to establish a rational and moral policy for their use as a deterrent, rather than to rely on a policy which placed supremacy over the Soviet power as its only goal.

In the space of five years, from 1950 to 1955, Murray spoke forth on atomic energy matters about twenty-two times. Many of these occasions were addresses before graduation classes at leading universities and others were before industrial and professional groups. For most of these speeches, the words that Murray used were crafted by Gerard, bringing out those thoughts that Gerard and Murray had developed in the many morning sessions at Murray's conference table. Extensive reading of current thinking was, of course, a part of the process but these two minds brought to the discussion a deep and sturdy faith in the Creator as a guidepost for the world to follow in seeking a way out of the dilemmas that it was facing.

On a typical Monday morning, Gerard would arrive at the office with a legal sized pad containing three or four pages of outline notes. This was to be the next Murray speech, the thrust of which had been sketched out in the previous week in those all-important conferences between them. Gerard would then proceed to flesh out the outline, dictating at a measured pace, never rephrasing, hardly pausing in the flow of words and thoughts. His thought process was so orderly that no second draft was needed before the proposed speech was presented to Murray. The next step, the "mark-up," was done in a "sit-down" with Murray and was the last step needed—so sure was his sense of the message and so unerring was his choice of the right words with which to say it. All of his writing was rich with references to history and quotations from philosophers and public figures of the past and present. Gerard seemed not to need a library to aid him in these well-chosen illustrations nor did he need, Heaven forbid, a Thesaurus; Gerard's memory was more than equal to the task of bringing these thoughts out as easily as everyday conversation.

In those speeches of Murray's, we can read the evolving thoughts of Gerard Smith on the challenge of the atomic age. In October 1953, in announcing Shippingport, the first solely civilian full-power nuclear reactor project, (an Eisenhower atoms-for-peace showcase) the keynote of Murray's speech was "far more important than war." In December 1953, at Marquette University Murray called for citizens' understanding of atomic energy issues with the theme "don't leave atomic energy to the experts." Re-reading the speeches today, we realize how timely they are even now—more than thirty years later. Many times and in many ways, Murray—and Gerard—pleaded with audiences to face up to the responsibility which the atom and its awesome power placed on the people of this country and of the world.

A theme often presented in these speeches is that there should be three bases to man's thought and action regarding the atom: *prayer* for God to give wisdom and help; *power*, which is necessary to act as the shield for equality in negotiating world peace; and *prudence* in the use of that power which comes from a knowledge of the facts of atomic power and an appreciation of its potential for universal destruction.

As the soundness of Gerard's thinking on nuclear policy became more widely known in Washington, it was inevitable that he was called to answer greater challenges, leading inexorably to the most significant chapter in his government career—the Arms Control and Disarmament Agency. Gerard, to the great regret of Tom Murray, left the Atomic Energy Commission with the equivalent of a doctorate in nuclear science not "honoris causa" but "laboris causa." He left behind him at all levels in the Atomic Energy Commission a network of friends and admirers who were honored to answer his call in later years.

William E. McGuirk, Jr.
Baltimore, Md.
October 1988

Foreign Policy Planning, 1957–1960

When Gerard Smith showed up to become head of the Policy Planning Staff of the State Department in 1957, to replace the much liked and respected Bob Bowie, who was returning to Harvard, none of the members knew him. Four years later, on January 19, when he walked out of his office into a cold snowy night, every member was deeply sorry that a change of administration had sent him back to private life. In those four years he had endeared himself to us—as a friend, a boss, and an effective counselor of the Secretary of State.

Anyone who has served as head of the Policy Planning Staff (known in State Department parlance as S/P) knows the importance of the relationship with the Secretary. Gerry had won the high regard of the Secretary by the quality, courage, and objectivity of his advice. Whether on defense, foreign aid, the Middle East, Berlin, or arms control, Gerry spoke up, with a view that was often different from the tepid consensus that the regional bureaus served up to Mr. Dulles. Sometimes

Gerry's views differed from those of the Secretary; but he did not draw back. Perhaps Mr. Dulles also liked Gerry because of their common concern with religious issues; in any event, they were both serious people and the Secretary listened to him with respect.

The best measure of their friendship and mutual respect was the humor that sometimes characterized their relationship. On one transatlantic flight, Gerry slept a good deal longer than Mr. Dulles and awoke to find the Secretary hard at work on his legal-sized yellow pad. Looking up to see Gerry descending from his berth, Mr. Dulles commented: "Ah, here comes the indispensable man."

The members of S/P liked the fact that their chief enjoyed ready access to the Secretary. This meant that their work mattered; it might affect U.S. policy. (In those days, believe it or not, the Secretary of State actually had something to do with making policy.) But the members of S/P liked Gerry for other reasons: He was interested in getting the best out of each of us; he encouraged us to develop our own ideas; he found time to talk about these items with us—individually and in frequent meetings of the staff. When he had meetings with the Secretary he would ask the relevant S/P expert to join him, whenever that was possible. He made us feel as though we were members of his team.

All this was coupled with a human touch. He invited us and our wives to dinner with his family and guests, and to outdoor parties on the Eastern Shore. When former members of the Staff were in town, they were invited to tell the Staff about their experiences. When Bill Leonhart returned from a spell at the British Imperial Defense College and told us the British were grieved to find that the registration numbers of revolvers retrieved from IRA terrorists in Ireland could be traced to the New York Police Department, Gerry observed: "I always knew no good would come from letting Italians into the Force."

Which brings me to something you've probably gathered by now: Gerry has a great sense of humor. He is a rather shy and reserved man, so it took a while for his humor to show. But when it did, we all enjoyed it. Not even the persons and causes closest to him were exempt. Two examples remain in my memory:

- His telling us about the southern Italian Cardinal who complained that he couldn't understand all this ecumenism . . . *he* had never met any Protestants.
- The story about the Irish truckdriver on trial in a British court, defended by the great barrister E.F. Smith, whom the judge haughtily asked, "Surely your client is familiar with the principle 'de minimis non curabit lex'," only to receive the taut reply: "M'lord, in the County Cork they talk of little else."

But it would be misleading to let you think that Gerry was all fun and games. On the contrary, he was deadly serious. As Marquis Childs remarked in one of his columns, Gerry's policy views reflected his deep religious concerns. Nowhere was this more evident than in his intense preoccupation with nuclear weapons. Many of the chapters in this book reflect this preoccupation. As told in one of those chapters, he had seen a nuclear explosion; this left its mark on him for the rest of his life. The threat that nuclear weapons posed to the survival of the human race, which God created, was the central theme that gave unity to his work and thought. Whether as a Special Assistant for Atomic Energy to the Secretary, or head of S/P, or chief arms control officer of the U.S. Government, or an ambassador-at-large trying to restrain nuclear proliferation—this theme shaped and directed most everything he did.

In S/P it led him, as it had led Bob Bowie before him, to educate the Secretary about the awesome power of nuclear weapons. I can still see him sitting at the conference table in his office, surrounded by maps with large circles showing how far the effect of atomic weapons used to defend the offshore islands (remember Quemoy and Matsu?) would extend, and how many human beings they would kill and maim. Since the conventional defense of the islands was difficult, it was natural that attention should focus on defending them by nuclear weapons. Gerry made clear why this was unacceptable. Perhaps this helped to defuse the hysteria over these islands in the late 1950s.

His appreciation of the devastating effect of nuclear weapons led Gerry in two complementary directions:

On the one hand, he tried hard to bring the U.S. Government to understand the need to build up U.S. conventional forces. I can remember him coming back from a meeting with the Joint Chiefs of Staff, where he had been trying to educate them about the limitations of massive retaliation and the need for more versatile forces. It was hard work. But they and other people in the U.S. Government listened—partly because they knew that Gerry had the Secretary's ear, but partly because his knowledge and honest concern were so evident. It was clear that he knew what he was talking about—and that he was not promoting anything but the facts. Least of all was he trying to promote himself.

Throughout, Gerry pushed for sensible arms control. For a while, the leading actor in this drama was Governor Stassen; he soon lost his principal constituent, the Secretary of State, by not consulting him. After that, the responsibility was Mr. Dulles', and Gerry was his senior back-stopper. That was the day of comprehensive disarmament plans, and Gerry worked with Phil Farley, Ron Spiers, and Vince Baker in developing and refining such plans, which finally faded away in a miasma of numerical overall ceilings. Gerry also worked for a test ban, which had a more promising future. The Secretary's doubts about the Soviets and his concern about the Republican right wing (it's hard to say which worried him more) made him less than enthusiastic; the President seemed more interested. Gerry kept pushing; his patience, disinterestedness, and knowledge impressed them both.

This was the more true because it was clear to them, and others, that Gerry had no illusions about the U.S.S.R. He perceived the wide schism between our moral values and theirs. When summit meetings between the President and Mr. Khruschev were being planned, Gerry wrote one of the best memos of his term urging that the President do needed business with the Soviets without any false front of amity that might obscure the profound differences between the two systems. Tommy Thompson read this memo in draft and argued with him about the value of detente; Gerry could never get him to be very precise about that value. Gerry's memo went forward to the President; Mr. Khruschev's performance at the intended Paris Summit both proved him right and made the argument moot; there was no opportunity for the President to treat Mr. Khruschev like a comrade.

None of this deterred Gerry from seeking agreements with the U.S.S.R. On the contrary, he felt these efforts were more likely to prosper if made on a business-like basis. So he kept at it, then and later.

He proposed arms control in one other way. Together with his predecessor, Bob Bowie, Gerry played a leading role in putting forward an alternative to national nuclear forces—the so-called Multilateral Force or MLF. It didn't succeed, but the basic principles involved—that long-range European nuclear delivery systems should be under joint, and not national control—is still one whose fulfillment would enhance world safety. It is another example of Gerry's life-long dedication to finding means to reduce the threat to the world posed by nuclear weapons.

Gerry's interest extended to other areas. When the 1958 Middle East crisis erupted in the wake of revolt in Iraq, Gerry not only argued against allowing U.S. forces to be involved in the Lebanon fighting (Under Secretary Murphy's on-the-spot guidance finally kept them safely grouped about Coca-Cola machines at the airport); he also proposed a constructive response. Mr. Dulles was skeptical, then left town on business, leaving Gerry free to help draft a Presidential speech to the U.N. that proposed a peace-keeping force, an economic development plan, and negotiations for a comprehensive settlement that might include the U.S.S.R. The Secretary returned irate, his reaction held in check only by the President's evident pleasure at the widespread approval his speech had elicited.

Another crisis that involved him often and deeply was Berlin. It's hard to remember how important that issue seemed in the 1950s, when the Soviets pressed again and again to dispossess the allies from Berlin. Their basic contention was that the Western occupation rights had waned with time and that a new regime for West Berlin was overdue. Gerry and I spent a long summer at the Geneva Foreign Ministers' Conference in 1959 helping to disprove this argument and to conceive constructive counter-proposals. Gerry saw that in the end it was not a matter of details but of will—our will to stay in Berlin and to use force to assert that right if necessary. He was deeply concerned, as the Geneva meeting ended, that the U.S. might reach a Berlin agreement with a fixed time limit and that this would imply we were only temporary tenants. Fortunately, that intent, if it existed, vanished in the confusion of the failed Paris Summit.

Gerard had a balanced approach to the use of force. He avoided excesses of belligerence as vigorously as he warned against giving away the "store." When there were problems over our flights to Berlin, because the Soviets were using certain air corridors and altitudes for anti-aircraft "exercises" and U.S. Air Force officers said that they could not maintain needed flights without using these corridors at these altitudes, Gerry's fascination with technical detail surfaced. He was skeptical of the Air Force's argument; in the end, it became apparent that they could fly above the level proscribed by the U.S.S.R., and an unnecessary U.S.-Soviet confrontation was avoided.

I wouldn't want you to think that Gerry spent all his time in S/P dealing with the great issues of war and peace. Like Bob Bowie before him (there was great continuity from one to the other), he saw clearly that peace rested in the end on a working world order.

In 1958 Gerry invited Walt Rostow and Max Milliken, MIT professors who had written a prescient book on economic development, to discuss their views on foreign aid with S/P. He was impressed. There was then much concern over Castro; it seemed to Gerry that there was more to promoting freedom in Latin America than working against the Cuban ruler. He commissioned S/P member Hank Ramsey and others to come up with a plan for helping to meet Latin American economic development needs. The result was a proposal for $500 million in new foreign aid—devoted largely to social welfare purposes. The Secretary of the Treasury and the Director of the Budget Bureau took a dim view; the latter suggested to the President that we would be acting strangely in giving aid for Latin American education when we were not providing additional federal aid for U.S. teachers. The President was unimpressed by the connection; his favor and the Secretary of State's support carried the day. The forerunner of the Alliance for Progress was born.

So you can understand why, as January 19 neared its end and darkness descended on a snow-bound capital, all of us in S/P felt we were losing a wonderful boss. George Morgan, Gerry's deputy, expressed something of our feelings when he escorted the departing director of S/P from the State Department to his car. We all would have liked to be with him. When Walt Rostow succeeded Gerry soon afterward, he was good enough to find many occasions to show that he shared the staff's feelings of affection and respect for Gerry.

A few years later we were all vividly reminded of Gerry, when an idea that he and I had conceived came to pass. We had proposed in 1960 that a hot line should be established between the President and the leader of the U.S.S.R., to serve as a channel of communication when the risk of war by miscalculation might be high. Under Secretary Livy Merchant had his doubts, and the plan had died. In the Kennedy administration, attention turned again to this matter and the plan was accepted, partly due to the initiative of Carl Kaysen in the White House. Gerry has a letter from President Kennedy, thanking him for this service (vigorously protested by the editor of *Parade,* who claimed the credit for himself). Gerry's friends in S/P, and they remained his friends after he had left, were proud to be reminded of the days in which he had led them in serving the country and the cause of peace.

Henry Owen
Washington, D.C.
December 1988

The NATO Multilateral Force, 1960–64

"Who came up with this nutty idea?" asked Senator Scoop Jackson as he arrived, late, at a Congressional briefing on the NATO Multilateral Force. Proving yet again that lost causes have few friends, most of the opinions expressed since the MLF's demise share the senator's lack of enthusiasm. And yet, able people who worked hard and long on the project came to feel that the MLF, properly implemented, could make an important contribution to major goals of the U.S. and its allies. With due respect to Mr. Jackson's shade, while conceding that we can never know how the MLF would have actually worked out in practice and conceding, too, that competent experts differ both on the desirability of its objectives and on its ability to achieve them, it was not intrinsically a stupid idea.

The MLF was, rather, a revolutionary idea: to create a truly international force, differing from ordinary allied forces made up of national contingents in that no part of it would respond to the orders of any individual country. (Gerry Smith, meticulous as always, after some study announced that the MLF was not actually unique, that the League of Delos was an approximate precedent. However, since Athens, the strongest member of that League, ended up taking control of its forces and using them to attack recalcitrant smaller allies, it was held in the MLF team that the precedent would be unlikely to attract our European friends. Gerry was persuaded not to disseminate his findings.) To prevent any one country from controlling the force or any part of it, crews would be of mixed nationalities, with no more than 40% of any crew coming from one nation, and all operational directives to the force would come only from the allied command. To make the force significant enough to generate the needed political impact and justify the expense, it had to have meaningful nuclear weapon power; the force proposed was twenty-five ships, of good sea-keeping merchant ship design, each carrying 8 Polaris missiles for a total of 200 weapons capable of reaching the Soviet Union with multi-kiloton yields. Some three quarters of the ships could be kept continuously alert at sea.

It became evident, as the project developed, that the MLF would be in effect a little separate allied navy, that such a navy would support, and run it, and that a little allied navy department needed a little allied government, or some aspects of a little government, to direct it. None of the needed institutions or measures was simple or self-evident. The MLF, in brief, was not only revolutionary, it was complicated.

Further, the MLF had to be started from scratch, with no potent initial base of support, political or military, at home or abroad. To succeed, it had to develop that support within our own and at least some allied governments. Since participating nations would have to pay for the force and would get, at least initially, only very limited control over it and little or no strategic nuclear support beyond that already pledged by the U.S., it is understandable that most first positions, throughout the alliance, ranged from indifference to opposition. Throughout the MLF campaign, enthusiastic converts were hard to come by.

What objectives seemed to justify so elaborate and difficult a project? There were two fundamental desiderata that in the view of the MLF advocates the inter-allied force would advance. The first was the goal of European unity, or "European unity and Atlantic partnership," sought by many since the end of WWII. MLF supporters held that a truly international NATO military structure, probably developing over time in weight, in influence, and in "Europeanization," would add an influential political institution to the existing Western European economic organizations and be a major step toward European political integration.

The second major objective that the MLF aimed to work toward was a satisfactory nuclear weapon arrangement for Germany—"the German problem." Germany was loyal to her 1954 renunciation of nuclear weapon production. Her nuclear position was, however, not adequately equitable for the long term. Germany was making the largest European contribution to NATO forces; she lay on the front line of any NATO-Warsaw Pact hostilities; she was virtually certain to be destroyed in any nuclear exchange. She was keenly interested in nuclear support and control; her insistence on nuclear equity could not but grow. Clearly, her situation and contribution entitled her to some sort of participation in nuclear affairs. At the same time, neither the Soviet Union nor Germany's allies could contemplate a national German nuclear force. The MLF, it was argued, would be a vehicle for true German nuclear participation without the unacceptable independent German force.

These objectives, brought to a head by a number of nuclear problems that built up in NATO in the late fifties, triggered the MLF campaign. Gerry Smith planted the initial seed. In the spring of 1960, then head of State's Policy Planning Staff, Gerry had State contract with Robert Bowie, his predecessor in office, for a study of NATO defense needs. The Bowie Report, issued in summer 1960, among other recommendations proposed an internationally manned collective strategic sea force. This was the MLF's genesis. And Gerry's Policy Planning Staff then undertook the prosecution of the concept.

With the arrival of the Kennedy Administration, Gerry left the official team, but a number of MLF supporters had by then been developed and were well positioned in State. They kept the project moving. In May, 1961, they got support for "a NATO seaborne force, truly multilateral in ownership and control" into a Presidential speech in Ottawa, thus placing the collective force idea officially on the list of NATO's alternatives. And in April, 1962, a National Security Action Memo (NSAM 147), based on a paper masterfully drafted and cleared by Henry Owen, stated U.S. willingness to join an MLF, giving the project official status throughout the U.S. government.

With NSAM 147 Mr. Kennedy authorized internal exploratory studies of the MLF, and tentative discussions with allies. In April of '62, Undersecretary George Ball visited Bonn and talked MLF with Chancellor Adenauer. In June, Secretary Rusk carried the idea to Europe. At home, State, in addition to acting as the lead bureau on all aspects of the operation, got Gerry back on board by engaging him as a consultant to develop an MLF briefing for the allies. And the Defense Department, startled by the unanticipated possibility that the force might actually come into being and that Defense would be stuck with making the U.S. end of it work, directed the Navy to do a study of the feasibility—not the desirability, which was taken to be a political matter for State's judgement, but exclusively the feasibility—of the multilateral concept as outlined in the NSAM.

The Navy study, directed by Rear Admiral Michaelis, concluded that there were no insuperable obstacles, legal, operational, or human to the MLF. The basic conclusion was that the force was feasible. But the study had examined the almost absolute bureaucratic immobility of some allied agencies, notably the fighter plane spare parts consortium, and emphasized that the MLF would be feasible if, and only if, it were given continuing support, at high levels, by the participating nations. Whether that condition would have been met in practice was, and remains, an open question.

The Michaelis study gave a welcome boost to the MLF supporters in State, but introduced one new problem. The originators of the MLF idea had assumed that the force would be carried in nuclear-propelled submarines. The study strongly recommended using surface, merchant-type hulls, at least in the force's first incarnation, arguing that a good case could be made for the military merits of the surface configuration, and that operating the mixed force would be tricky enough without putting it in relatively sensitive submarines, driven by active, ticking, nuclear power reactors. (Nuclear weapons, by contrast, are inert until fired.) The State MLF team held, correctly, that prospective participants would regard the surface mode as a second class substitute and be substantially harder to convert to the idea. A sly attempt to get a decision for submarines, backfired, however. The MLF men arranged for Admiral Rickover, passionately opposed to the notion of foreigners in his submarines, to call on the President. Mr. Kennedy was briefed to listen politely, then tell Rick that the political requirement overrode his military argument. In the event, however, there was no record of the President having said anything whatever. Rick, a tough arguer, instead solidly convinced Mr. Kennedy that the submarines would not do. From then on, the surface configuration was the U.S. position.

In preparing his briefing for the allies, Gerry was constrained by other guidelines as well as the surface mode. He was specifically not to make a sales pitch for the MLF, only to outline a project which, if several allies wished to participate in setting it up, the U.S. would be prepared to join. Gerry's charge, basically, was to take soundings and report how much support for the MLF, if any, existed among the allies. He was punctilious in following his guidance, but it was obviously impossible to convey absolute U.S. neutrality. Since he was describing a U.S. idea, of which he was an official U.S.

messenger, our allies could not avoid concluding that we saw some merit in it; otherwise, why was Gerry making the trip? Inevitably, there was some feeling, in Europe and at home, that the U.S. was trying to pursue yet seem pursued.

The most difficult constraint, which hung over the project to the end, was the question of control: who could authorize the firing of the weapons? Here, the President insisted that he must either himself direct or at a minimum concur in the firing; either hold the firing key or at least a safety catch. No control system would be acceptable to the U.S. which would allow launching by one or several allies without his participation. This meant, of course, that the MLF's weapons would fire only when U.S. strategic forces were firing; in operational reality they would be just a part, some five percent, of the U.S. retaliatory forces, and if other countries were also to hold safety catches, the MLF would be the least responsive five percent. Such an arrangement would not appear to contribute much to integrating Europe, nor, from a European nationalist standpoint, to justifying the expense. To ease this point, U.S. MLF briefers and negotiators, then and later, were authorized to suggest, informally, that the control question might be reexamined at some unspecified later time, perhaps when European integration had produced a single executive able to act for Europe.

So charged and constrained, Gerry set out in October 1962 for the NATO Council in Paris, with Rear Admiral John M. Lee, from Defense, to speak to the cost and operational feasibility of the concept and to defend the surface mode, and with Abram Chayes, the State Department Legal Advisor, to argue its legality. On arrival, Gerry's team ran head on into another problem: the Cuban missile crisis. Chayes was recalled to Washington the day before the council briefing. Lee lasted until just after the initial briefing before he was jerked home. Gerry was left single-handed, feeling, as he has said, somewhat lonely, though he was fully capable of making the entire case and did so handsomely.

At the end of Gerry's trip, as 1962 drew to a close, European reaction could be scored about as follows: West Germany unequivocally pro, Belgium generally favorable, Italy nearer the fence but probably supportive in the long run, the U.K. reserved and skeptical but not adamantly opposed, and France completely negative. In the U.S., there were advocates and antis, but at the top of the hierarchies Rusk and McNamara were skeptical-to-negative, and the President not yet seriously seized of the problem. The campaign had not developed a wave of support, and it largely lost its momentum. It was at this point, in December 1962, that external events—the Skybolt decision, the Nassau meeting, and the DeGaulle reactions—pushed the MLF back into the front line of U.S. diplomacy.

In very sketchy outline: Skybolt, a U.S. air-launched ballistic missile of strategic range, was cancelled for lack of technical promise. The U.K., which had an agreement to buy Skybolts, was counting on them to maintain the declining capacity of its independent nuclear force; the cancellation undermined the force's entire future. At the long-scheduled U.S.-U.K. summit in Nassau on 19 December, Prime Minister Macmillan passionately demanded a new weapon agreement committing the U.S.

to the independent British force. President Kennedy yielded to the pressure and agreed to provide Polaris missiles to the British.

A little rhetoric was written into the communique which gave some color to the subsequent claim that the Nassau communique affirmed the principle of a mixed manned collective force. The reality was, however, that the support Nassau gave a national nuclear force flatly contradicted the MLF design for European unity. And President DeGaulle promptly drove home the point. He rejected an equivalent Polaris offer for France without bothering with negotiations; he refused to participate in any NATO nuclear force; he terminated EEC negotiations with Britain; he signed a treaty of cooperation with the Germans. Unified deterrence, Atlantic partnership, and European integration were suddenly in eclipse, and a split between the continental powers and the Anglo-Saxons seemed to threaten.

At this point, the MLF was the only game in town for the U.S. to use to undo the damage resulting from Nassau and DeGaulle. It offered a unifying NATO-wide structure, one which might ultimately absorb the French and British nuclear forces. It was a tool for much-needed broadening of German-U.S. nuclear relations. It was the only proposal that did not have to be improvised on the spot; the Smith-Lee briefing had familiarized all NATO nations with its outline. A group of effective advocates were well stationed in the State Department. As a result, the MLF became the objective of a major U.S. diplomatic campaign through 1963 and 1964 that very nearly brought the force into being.

The process commenced with another critically important paper by Henry Owen. This reached the President by the end of January 1963, and obtained his approval for proceeding with negotiations within NATO to establish the force, without insisting on the previous requirements that the allies pay the entire bill or that they first achieve conventional force goals. Within State, a special MLF office was established. Livingston Merchant, an MLF enthusiast, was put in charge, and appointed Special Ambassador for the negotiations. Gerry Smith was assigned as the State senior representative in the office; Defense sent over Squidge Lee and a few staff officers for full-time MLF duty. This office took over the day to day business of the MLF. Support came from the Policy Planning Staff, with Walt Rostow, Henry Owen, and input from Robert Bowie. Robert Schaetzel, Deputy Assistant Secretary in the European bureau, kept that organization on the right track. High level activism was provided by Under Secretary George Ball. In the White House, McGeorge Bundy was an advocate. In brief, the MLF supporters, in the aftermath of Nassau, established a network of able and well-positioned supporters, an organizational setup that gave them bureaucratic control, and a mandate to go ahead.

Step one for the reinvigorated team was to get a Presidential directive for European meetings. In two February White House meetings, with Merchant, Smith, and Lee present, Mr. Kennedy made clear that he remained skeptical about the success of the MLF proposal, that he did not want the U.S. tied too closely to it in case it failed, and that he did not propose to relinquish an American veto over firing. His concerns were inflamed further after learning that the MLF would require

U.S. nuclear legislation, and that legislation would be hard to get. However, the MLF case was ably made by Livy Merchant and by Gerry. At the end, the President had been brought to the point of issuing instructions to the team, directing it to investigate the possibility of an agreement on the MLF, and to negotiate a preliminary agreement if two or more governments proved interested. The President, though not without reservations, was now adequately aboard and a committed team was cleared to carry the MLF proposal to Europe essentially as they had designed it.

In March the Merchant missionaries consequently set forth for *partibus infidelium,* and took the word to the NATO Council and the capitals of U.K., F.R.G., Netherlands, Belgium, Italy, Greece, and Turkey. Receptions everywhere were polite but qualified. The Foreign Minister set the tone of the U.K.'s reception with an impressive flight into the subjunctive: "Ah, Livy," he said heartily, "It's good to see you again, and I want you to know straight away that Her Majesty's government is not necessarily opposed to your proposal, if, indeed, it should prove to be practical after further study." Other countries were more forthcoming in various degrees, with the Germans, again, most receptive.

On return to Washington, somewhat the worse for travel, the team felt it had found enough interest to recommend that the U.S. move hard for a preliminary agreement to establish the force, to be initialed during the President's trip to Europe in June. The White House decided that that went beyond the interest shown in Europe. The President did agree to write favorably to Chancellor Adenauer, but he held further decisions until the Chancellor replied. During the interval, what came to be known as the "MLF Cabal" kept the pot boiling. Walt Rostow and Bob Schaetzel wrote papers. Assistant Secretary Tyler visited Bonn and Rome. Vice Chief of Naval Operations Admiral Claude Ricketts went to Bonn and talked effectively with the German military. The JCS emitted a decision that the MLF was feasible and would be militarily useful. The AEC certified that the physical control arrangements contemplated could be made adequately secure. And at the end of April, Adenauer replied, accepting the President's requirements on control and surface configuration, and recommending a preliminary agreement while the latter was in Europe.

Adenauer's reply essentially met Mr. Kennedy's previous conditions, and the situation looked bright for action during his trip. But it turned out that the President was still unready for an early initiative. In May, he said that a German-U.S. MLF was not broad enough to be politically viable, and set the further requirement that Britain's participation be assured and that the U.K. agree to carry at least ten percent of the cost.

The Cabal made a full scale try with a flurry of diplomatic activity, but time was short and external events deflected international interest. There was no chance for agreement during the President's trip in June, and his references to the project while in Europe were confined to a general affirmation of the MLF concept and its logic, and an agreement to conduct quiet discussions without a deadline or target date. With that, the MLF project came off the boil. In the White House, there was a feeling that if the Europeans did not actively want the force, the initiative should probably be allowed to die.

The MLF team squared its shoulders and kept going. The approved quiet discussions were set up in Paris, consisting of the NATO ambassadors of the seven interested countries, known as the Working Group, charged with working out the legal and political structure of the force and preparing a draft treaty. A sub-group was established in Washington to discuss technical military matters: ship design, command arrangements, operations, financial arrangements, etc. In addition, a mixed-manned demonstration was undertaken. The missile destroyer USS BIDDLE, subsequently named RICKETTS after the late, pro-MLF admiral, was given an international crew and, so manned, participated in NATO exercises and port visits in the ports of the Working Group. These efforts kept the project alive, but not advancing. MLF needed a political deadline, to force top-level consideration and decision.

Major events made early decision impossible. In the fall of 1963, President Kennedy's assassination brought President Johnson to the White House. The U.K. shifted Tory Prime Ministers, from Macmillan to Douglas-Home. In the Federal Republic, Erhard replaced Adenauer as Chancellor. It was clearly unrealistic to expect agreement on a major NATO initiative until all three had time to get settled. What that meant was that the earliest feasible deadline was not until late in 1964, after the British election in October and the U.S. election in November.

The MLF organization went to work to bring that deadline to pass. In April of 1964 Gerry took over the

MLF Negotiating team from the retiring Livy Merchant, and almost immediately succeeded in getting the date set. At an April meeting in the White House, Gerry, George Ball, and Thomas Finletter, our Ambassador to NATO and a long-time MLF activist, reviewed the state of the project and made the case for the force. They left with Presidential authorization to go after an agreement by the end of the year. When Chancellor Erhard visited the President in June, the matter was nailed down publicly; the communique read. ". . . the proposed multi-lateral force would make a significant contribution to the political and military strength of NATO and . . . efforts should be continued to ready an agreement for signature by the end of the year." This committed Mr. Johnson personally, over and above MLF statements Mr. Kennedy had made. For Chancellor Erhard, the commitment involved making a substantial political investment; to support the MLF meant bucking French and domestic opposition; he needed the MLF out of the way comfortably ahead of his elections. The MLF had a deadline.

And soon, the project had a decision point. After the British election in October 1964 brought in Harold Wilson as the new Prime Minister and the American election in November returned Lyndon Johnson with his own mandate, a meeting of the two was clearly called for. It was set for early December. British participation was critical for the MLF and the MLF was critical for the Germans; inevitably the MLF would be on the Johnson-Wilson agenda. In brief, the MLF project was on the way to another summit, this one decisive, and the administration in Washington was faced at the highest level with deciding what the President would say about it to the Prime Minister when they met.

The White House set up a committee, co-chaired by Ball and Bundy, with McNamara and Rusk, started by Richard Neustadt, to recommend a position to the President. They made a careful and fresh review of the entire project, including coverage of the opposition, at home and in Europe (interestingly, some of that opposition came from Gerry's subsequent command, the Arms Control Agency, which feared that the MLF would interfere with the progress of the Non-Proliferation Treaty). The heart of the Committee's report was the recommendation that the President hold firm on the surface fleet and insist on British participation in it. Bundy added the judgement that Johnson was strong in the situation and Wilson weak; the Prime Minister could be made to agree. Later soundings confirmed this opinion. Had the President accepted the recommendation, the MLF would almost certainly have come into being.

The President, however, reached a contrary decision at two meetings with his principal advisors on 5 and 6 December. He appeared to accept the basic arguments of the MLF advocates, but based his conclusion on the internal politics of the project, in European countries as well as the U.S. He understood and sympathized with Wilson's problems. He brought out the opposition of the French, the doubtfulness of the Italians, the pressures on Erhard. He concluded that he could not take the MLF to the Congress as response to European pressures. He was aware by personal contact of powerful opposition in the Congress, and did not feel he had a sufficient mandate to overcome it. He learned, apparently for the first time of President Kennedy's reservations, and shared them. He was not moved by the argument that he was personally committed by the Erhard-Johnson communique, nor that Erhard, having committed himself to the MLF against strong opposition, would be badly embarrassed at home by a U.S. reversal.

Each of his chief advisors—Rusk, McNamara, Ball—supported the Committee's recommendation to go ahead. President Johnson, however, held firm and consequently the nays had it. As Wellington said of Waterloo, it was "a damned nice thing—the nearest run thing you ever saw in your life," but at the end of the 6 December meeting, the MLF was dead, even though formal obsequies dragged on for a time.

In the immediate follow-up, Gerry Smith resigned, the MLF office in State was abolished, and all substantive action ceased.

In summary, Gerry was at the heart of the MLF project from the moment he started it in 1960 by commissioning the Bowie Report until, as Head of the MLF Negotiating Office, he built its input to the December 1964 meetings where the MLF was sunk. From start to finish, Gerry's contribution was fundamental to the development of the concept, to its adjustments to meet political and diplomatic necessities, to its presentation and defense in the U.S. and Europe. Throughout, he was wise, dedicated, sensitive, scrupulous. It was rewarding and illuminating to work with him; we who did so were moved not only to admiration but to affection as well.

At the end, to be sure, the MLF met defeat, but other captains whose reputations have endured also finally met their Waterloos or their Appomattoxes. Gerry's campaign, like theirs, was hard and skillfully fought. And it was pressed home to the very last: Gerry did not flag nor fail. He endured unto the end.

Admiral John Marshall Lee
St. Petersburg, Florida
October 1988

Director of ACDA, 1969–1972

Gerry Smith became Director of the Arms Control and Disarmament Agency (ACDA) in January of 1969, at a time when U.S.-Soviet strategic arms issues were clearly becoming a cutting edge of U.S.-Soviet relations. The abortive 1968 effort of the Johnson Administration to get strategic arms limitation talks (SALT) started had been cut short by the Soviet invasion of Czechoslovakia, but the domestic debate over ABMs had intensified. In appointing Gerry as Director, President Nixon had said the role and status of ACDA was to be upgraded. Behind Mr. Nixon's public stance, however, lay a distrust of the Agency, though not of Gerry, that was to make the new Director's task more difficult during the next four years. The ACDA, while small in numbers, with less than 200 professionals, a budget of around ten million dollars and professional depth in the strategic area not yet fully developed, had by the time Gerry assumed command become an effective and, in most quarters, respected agency.

As an agency of limited resources in competition with the established Washington bureaucracies, ACDA was never destined to match the rather exalted responsibilities set for it in its 1961 creating legislation. Nor was a Director ever to be fully regarded by chief executives as, in the words of the 1961 act, "the principal advisor to the Secretary of State, the National Security Council and the President on arms control matters." Nevertheless, under Gerry Smith, ACDA was to reach its highest level of professional competency and influence, a level not since matched.

As Director, Gerry brought with him the perspective and wisdom acquired through many years of official involvement in arms control and national security issues. He also brought a personal view about nuclear weapons that put a steely determination in his efforts to control the arms competition. It was a long-held view; he had once told Secretary of State Dulles in the years before the limited test ban treaty banned atmospheric testing that watching a nuclear explosion was like "having a look at Hell." It also set him apart from most of his senior government colleagues during his tenure at ACDA; I remember a private comment he made after one particularly nonproductive NSC session: "The trouble with that meeting was that I was the only one in the room that has ever seen one of those things go off."

Curiously, and probably unbeknownst to him, Gerry was initially viewed with considerable apprehension by much of the professional staff of ACDA. The principal task and achievement of the Agency during the previous three years had been negotiation of the Nonproliferation Treaty (NPT), and the main initial barrier to achieving the treaty had been the ultimately-abandoned project for creation of a Multilateral Nuclear Force with our European allies. Gerry's role as head of the MLF task force in the State Department was the basis for this initial staff concern.

It was a measure of the man how quickly this attitude changed. First came the perception of the personal qualities of the new leader: his dignity, seriousness and above all a sense that he was, in the words of one staff member, "a true gentleman of quality." These things counted with the troops, particularly when, by the nature of the task, they were often viewed by others in the bureaucracy as bucking the prevailing currents and interests. And it also soon became clear to the staff that Gerry was prepared to speak out boldly within the government on arms control issues. Finally, there was the nature of his appointments; and in almost all cases the White House accepted his choices.

It was reflective of his bipartisan approach that he privately told most of the senior incumbent political appointees that they would be welcome to stay on. They graciously declined. (It seems doubtful that White House approval would have been forthcoming had they accepted.) Adrian Fisher, the outgoing Acting Director, did serve for an initial period in 1969 as the U.S. Representative to the multilateral talks in Geneva.

As his Deputy Director, Gerry chose Philip Farley, a former colleague and one of the Foreign Service's most respected national security professionals. Not only would Farley be an invaluable aide to Gerry during the "Washington phases" of his tenure at ACDA, but to Farley would fall much of the administrative responsibility for the Agency as well as responsibility for ACDA's role in "backstopping" the SALT and other negotiations during the periods when Gerry was abroad as head of the SALT Delegation.

Other key appointments to head Agency bureaus included Spurgeon Keeny for Science and Technology, James Leonard for International Relations and Admiral John M. Lee for Weapons Evaluation and Control. Keeny brought lengthy experience in arms control from his many years in the White House in both the Science Advisor's office and that of the National Security Advisor; his bureau was to carry the Agency's responsibility for SALT. Leonard was a senior Foreign Service Officer, whose skills would be demonstrated in the negotiations of the Seabed Treaty and the Bacteriological Weapons Convention. Admiral Lee was an associate of Gerry during the MLF project and was a naval officer with both high humor and broad intellectual scope.

Throughout the Agency, Gerry sought to maintain professional continuity by retaining such people as Sidney Graybeal as Keeny's deputy and Charles Van Doren as Deputy General Counsel. I was surprised when Gerry created the position of Counselor for me, since I had been Fisher's assistant and the most recent of our limited prior contacts had involved an argument in a public forum over the MLF.

The initial period at ACDA was a busy one. In addition to filling out and strengthening the Agency's resources, there were the interagency preparations for SALT, as well as the ongoing multilateral negotiations in Geneva at the Conference of the Committee on Disarmament (CCD). Gerry's SALT responsibilities precluded delegation participation in the CCD deliberations. He did decide, however, that he should attend the first CCD session in 1969 as a manifestation of interest in its activities. The one noteworthy event of this brief visit to Geneva afforded an early insight into Gerry's interest in gathering all the facts on an issue; it is also an interesting historical footnote to Gerry's participation years later in the effort of the "gang of four" (as Secretary Haig termed Gerry, McGeorge Bundy, George Kennan and Robert McNamara) to open up discussion within the establishment of the no-first-use issue with their *Foreign Affairs* article, "Nuclear Weapons and the Atlantic Alliance."

Soviet Representative Roshchin asked for an early private meeting with Gerry in which he raised the issue of a no-first-use agreement for nuclear weapons. Saying the Soviets thought progress on this question could be made step-by-step, he made references to informal discussions between some U.S. and Soviet officials during previous CCD sessions. Gerry limited his response to noting current U.S. views on this issue which made prospects for useful discussion dubious, saying he would, of course, report Roshchin's comments. As we left the meeting, Gerry said he had never heard of some of Roshchin's references, and when told there was a richer history of U.S.-Soviet exchanges on this subject than was generally known, he tasked me with the job of pulling the record together. The resulting product was a highly classified document which Gerry sent to the Secretary of State and the President with a comment that it was important that senior officials be aware of all the facts on this issue.

On the substance of the no-first-use question, Gerry

was meticulous in staying within the guidelines of the government's strongly negative views, whatever his personal thoughts. In part it was probably the result of his practical judgment that with the difficulties ahead on SALT, which he regarded as the primary task of the moment, his political capital had to be used with a sense of priorities. Beyond this, it reflected his view of the proper role of an official, stay within your instructions, a view that had been reinforced by the lesson learned years earlier in observing at close hand the troubles that had befallen Harold Stassen when he had exceeded his instructions.

Gerry was prepared to fight hard within the government for positions he believed in, and in the SALT and multilateral negotiations to explore carefully possible openings and to shrug off some of the skittish middle levels of the bureaucracy; but on known positions of the government, he never let the drawbridge down. The fact that this was not always recognized by the White House was probably galling for him; but it did not change his views or his conduct regarding the proper role of officials.

The account of the SALT negotiations and Gerry's role as head of the SALT Delegation is contained in another section of this book. However, a brief mention might be permitted here of two of Gerry's early Washington efforts on SALT prior to the openings of negotiations in late 1969. They were both unsuccessful; in retrospect, the rejection of Gerry's advice represented serious missed opportunities in strategic arms control.

The first concerned his efforts, with some support from the State Department, to get a moratorium on testing of MIRVs (multiple independently targetable reentry vehicles) in order to prevent the developing technology in this program, where the U.S. lead was substantial, from outdistancing the slower pace of the upcoming negotiations. His advice was rejected, and—to shorten and simplify a tragic and complicated story—the forces behind the momentum of the U.S. MIRV program became too strong to allow a serious U.S. effort to halt MIRVs; this ultimately resulted in a quantum jump in strategic warheads on both sides from roughly 2,000 up to 11,000.

A related subsequent effort before the SALT negotiations got underway was Gerry's advocacy of a negotiating proposal of his staff that was called Stop Where We Are or SWWA. The SWWA proposal included a mutual undertaking to halt MIRV testing and all strategic systems construction programs; it was an arrangement that Gerry believed not only would be in America's security interest while in effect, but would also preserve the President's options during future efforts to negotiate strategic arms limitations. It does not require strategic computer runs to reach the common sense conclusion of how much better off we would be today—even compared to the hoped-for START agreement—if SWWA had been successfully pursued. These episodes also demonstrated that, however careful as a negotiator in staying within his instruction, Gerry was prepared to be an advocate of bold proposals, a characteristic so often lacking among senior officials, fearful of being labeled "extreme."

As both ACDA Director and head of the SALT Delegation, Gerry's plate was always full: constant shuttling across the Atlantic for SALT sessions; the month-long sessions themselves, with hotel-room living, which Gerry considered one of the lower forms of lifestyle; the interagency SALT studies and debates in Washington during recesses; the multilateral arms negotiations; and the administrative duties of running and strengthening ACDA. Through all this, it was clear that he derived much strength and comfort from his gracious wife, Bernice.

Gerry always seemed to be working. This included what the rest of us called lunchtime. No seventh floor State Department executive dining room for him. He ate and worked at his desk. This period was often used to improve his Russian, a language with which he had earlier acquired a limited familiarity. This labor–and working on Russian over a sandwich is nothing if not labor–was of some help to him in the SALT negotiations. It was also singularly helpful on one occasion early in SALT, when there was still a certain stiffness between delegations, in establishing a personal rapport between Gerry and the Soviet Delegation.

After some hesitation, Gerry decided it might help the atmosphere to deliver a U.S. statement in Russian, and asked Bill Krimer, the U.S. interpreter, to check a Russian language version of the U.S. statement. When, at the meeting, he began his statement in Russian, a perplexed look appeared on Soviet Representative Semenov's face. As he continued, the Russians began to look questioningly at each other. Then smiles appeared, and spread from Semenov to academician Shchukin to General Ogarkov and on down the table. When Gerry finished his statement by saying he would "now like Mr. Krimer to interpret my statement for the benefit of the

U.S. Delegation," all stiffness evaporated. One could sense the personal rapport that had been created by that gesture. Gerry's lunchtime labor at his desk in Washington had produced a small diplomatic gem.

The staff meetings at ACDA were never stiff or pro forma. Candid exchanges were not limited to the "small staff" meetings but were equally free at larger staff discussions. Gerry sought, usually unsuccessfully, to have the same open discussions at NSC meetings. It was typical of his work attitude that for the many interagency studies that were supposed to be the basis for such meetings, Gerry read carefully the full reports as well as the "executive summaries," a practice that set him apart from most of his NSC colleagues.

His intellectual curiosity and attention to papers that crossed his desk were both an inspiration to his staff and, at times, almost overwhelming. At his desk Gerry kept a dictaphone constantly at the ready. He often wanted to know more than what was in a paper, or why such and such had not been considered, or "what if?" The informal notes, which became known as "Smithograms," that went out to various agency officers made everyone aware the boss was reading their work carefully, and in numerous cases prompted new thinking.

However, the staff never treats questions from an agency head lightly, however informal they are. As the activity of the staff increased, so too did the Smithograms, and a staff-perceived need for some relief developed. I was talked into approaching Gerry. He was furious with me. He felt that I and the rest of his staff should be pleased that he took such close interest in the staff's work and that, in any event, it was important that a Director have full knowledge of facts and reasoning and that he stimulate the thinking of his staff. No give was indicated in his reaction, but in subsequent weeks the number of his notes declined. Fortunately, Smithograms continued to be an endearing and sharpening feature of his stewardship.

While his principal preoccupation was SALT, Gerry's interest and involvement in the multilateral negotiations did not slacken. He relied heavily on Farley for agency direction in these multilateral efforts, especially during his lengthy periods in Helsinki and Vienna. Even while abroad, he assigned a delegation officer with the special task of keeping him informed of on-going developments in the multilateral negotiations. And there was a great deal of activity in the multilateral area.

While the Nixon Administration maintained the U.S. commitment to a comprehensive nuclear test ban, it became clear that no priority effort would be made to resolve the issues standing in the way of an agreement. Gerry accepted this. Among other reasons, there was too much else going on to get the test ban to the front burner of decision-making and too much capital to be lost in trying to do so. But under ACDA sponsorship, rather vigorous research and field test efforts to simplify and improve verification capabilities continued, especially in the area of seismic detection and identification of underground tests and including improvement of unmanned seismic stations or "black boxes."

One area of arms control that did not even produce negotiations should be mentioned. It was a so-called "cut-off" proposal to halt production of fissionable-material for weapons purposes, a proposal that had originated in the Eisenhower Administration. There had always been strong opposition to this concept in the Pentagon. Through his efforts, Gerry managed to maintain U.S. advocacy of the proposal during the Nixon Administration. In addition, approval was obtained to have the U.S. propose that, rather than the adversary type inspection for this measure that the U.S. had previously called for, inspection of permitted peaceful nuclear facilities would be by the International Atomic Energy Agency in the manner that the agency applies inspection for non-nuclear parties to the NPT. It is one of the many ironies of arms control that when, under Gorbachev, the Soviets came around to acceptance of the cut-off concept, U.S. officials seemed to have forgotten it. Gerry, at least, kept it alive on his watch.

ACDA also carried responsibilities for the Mutual and Balanced Force Reduction (MBFR) talks for Europe. These ultimately fruitless negotiations did not begin until after Gerry left ACDA, but the Agency's work in the interagency and NATO studies during his tenure was yet another ball he had to keep in the air.

Success in one multilateral effort where ACDA had primary responsibility came relatively early. Negotiations had begun in 1967 on a treaty, the basic provisions of which would ban placement of nuclear weapons and their launch facilities on the seabed and ocean floor. In 1969, the Soviets abandoned their position that all military uses of the seabeds, including submarine surveillance systems, be banned. This made possible the submission in Geneva of a joint U.S.-Soviet draft treaty. The Seabed Treaty embodied a simple concept of "non-armament," to preclude an area from arms competition before one had begun. The negotiation of a final draft, however, was not easy, for complicated extraneous issues stemming from concerns about precedents for the law-of-the-seas negotiations had to be resolved through negotiations in the CCD and in the U.N. in New York. In ceremonies in Washington, Moscow, and London in February 1971, the Treaty was signed by over sixty countries.

A second multilateral agreement was achieved by 1972 with the signing of the Bacteriological Weapons Convention. Negotiations leading to this accord were greatly stimulated early on by President Nixon's actions. After interagency review of the issues, the President, in November of 1969, announced that he was submitting to the Senate for ratification action the 1925 Geneva Protocol. The Protocol, prohibiting the use of poison gas and biological warfare, had been the product of a U.S. initiative, but for a variety of reasons, including, in later years, fear of a possible precedent for nuclear weapons, the U.S. had never ratified it. The President further stated that the U.S. would dispose of existing stocks of bacteriological weapons, would confine further research to defensive measures such as immunization, and that the government would associate itself with the principles of a British draft convention to ban biological weapons (leaving to a later date the more complicated question of a ban on chemical weapons). This bold unilateral step surprised most of the bureaucracy and, contrary to the predictions of skeptics who argue that unilateral steps remove incentives for other parties to bargain, it ultimately led to the 1972 Convention.

The negotiations, however, were time-consuming and

complicated. It was not until the spring of 1971 that the Soviets abandoned their insistence that chemical weapons also be included; a joint U.S.-Soviet draft was then worked out as the basis for yet further multilateral negotiations in the CCD. The Convention was finally opened for signatures in April of the following year, one month before the signing of the SALT I agreements.

The Seabed Treaty and Bacteriological Weapons Convention were not the only agreements reached during the busy period prior to the Moscow Summit and the SALT I accords. While negotiated by the SALT delegations, two other agreements were reached that were not limitations on strategic arms and for which Gerry and his agency carried significant responsibility. With the aim of reducing the risk of accidental nuclear war, technical working groups were established by the delegations. The first group explored ways of improving the reliability and survivability of the U.S.-Soviet Direct Communication Link established in 1963. The second discussed arrangements for an exchange of information to reduce uncertainties and misunderstandings in case of a nuclear accident or certain other events.

In the spring of 1971, the outlines of agreements in both areas were reached; by the end of the summer, Gerry and Semenov were able to initial the accords in Helsinki, and the formal signing was accomplished in Washington in September.

The "Hot Line" upgrade agreement included various measures to improve the original agreement, the most important of which was to provide for the addition of two satellite communication circuits, one U.S. and one Soviet, to provide more secure communication than that afforded by the cable and radio circuits of the 1963 accord. The agreement on measures to reduce the risk of war spelled out obligations and procedures for immediate notification in such cases as accidental or unauthorized use of nuclear weapons or detection by warning systems of unidentified objects; it also provided for advance notification of a specified type of missile test. These agreements were small but not unimportant steps toward a safer world, and Gerry derived considerable satisfaction from their achievement.

If all this were not enough to keep Gerry and his agency busy, there were the continuing efforts to complete and implement arrangements relating to the Nonproliferation Treaty, for much of the procedure for its implementation still had to be worked out. Various technical research and development programs to assist IAEA verification were conducted by ACDA, in coordination with that organization, other U.S. agencies and other countries. Moreover, the negotiation of specific safeguard arrangements, and particularly the special 1972 inspection arrangements between the IAEA and EURATOM, the Western European nuclear agency, occupied a significant portion of ACDA's resources.

In another nonproliferation area, the U.S., in 1971, ratified Protocol II of the 1967 Tiatelolco Treaty prohibiting nuclear weapons in Latin America. This protocol committed the U.S. to certain limitations on the use of nuclear weapons against parties to that treaty.

Under Gerry's leadership, ACDA worked on the problem of the traffic in transfers of conventional arms. The Agency also sponsored and published significant studies and other materials on world military expenditures and on the impact of military spending reductions on the nation's economy.

With all this activity, together with the significant strengthening of ACDA's technical competence in the strategic area, the addition of the increasing cost of the SALT negotiations imposed severe strains on the Agency's budget. Despite its modest budget, ACDA carried the burden of most of the government's expenditures for those negotiations, including such unlikely items as paying for the costs of the special Marine security detachments and equipment which traveled back and forth between Helsinki and Vienna.

At one point, Gerry sought some assistance by penning an informal note to Secretary of Defense Laird in which he reviewed the financial strain imposed by the negotiations and suggested that the Secretary, with his larger budget, might be able to come up with a little supporting money—as I recall, the figure was $100,000. Mr. Laird's reply indicated an understanding of Gerry's plight, but, in effect, said that the Department's tight budget requirements were such that no funds could be spared. Gerry was not really surprised by the response, but he did utter some private observations concerning priorities in the allocation of the taxpayers' dollars.

Gerry's relationships with senior officials of the Nixon Administration quite naturally varied. Policy differences with Mr. Laird precluded a close working relationship, but their differences never seemed to drift into personal animosity. This was even more so with General Wheeler, the JCS Chairman, with whom Gerry had worked prior to his ACDA assignment. In both cases, the open and respectful relationships that Gerry maintained on the

SALT Delegation with Paul Nitze and General Royal Allison, the Department of Defense and JCS SALT Representatives, helped keep Washington policy differences generally devoid of personal rancor.

While a separate agency, ACDA had always had a special and close working relationship with the Secretary of State; he was, for example, always informed of any direct independent communications the Director had with the President. The close similarity of views on SALT and other arms control issues that Gerry shared with Secretary William Rogers, and the respect they shared for each other as individuals, assured continuation of such a relationship.

The more critical relationship, that with Henry Kissinger, was another matter. In his book, *Doubletalk*, Gerry has recounted the interplay of that relationship, and little of value can be added here. Gerry respected Kissinger's intelligence and capacity for hard work, but he found it increasingly difficult to accept the deception that marked some of Kissinger's dealing with him as head of the SALT Delegation. Only Gerry really knows at what point he decided he would resign once the SALT agreements had been reached. I suspect that it came in the spring of 1972. In an exchange of back channel messages, Kissinger had promised Gerry that he would keep him "fully informed" of any parallel SALT negotiations that might occur. Shortly after this assurance, Kissinger secretly flew to Moscow where negotiation of what was to become the missile submarine protocol to the Interim Agreement took place. In contrast to Semenov, who had been recalled to Moscow for the meetings, Gerry was informed of Kissinger's trip after the fact and a day before the public announcement. This episode, together with Nixon's angry retort of "bullshit" to Gerry's effort at a subsequent NSC meeting to improve the formulation resulting from the Moscow meetings, a retort Gerry felt raised questions about the President's attitude toward him, were probably the turning points.

Until this period, Gerry's relations with President Nixon seemed generally good, if rather formal, particularly in light of the President's suspicion of ACDA, itself. Nixon had more than once accused ACDA of being the source of leaks to the press, charges that Gerry was certain were inspired rather than true. In fact, ACDA officials leaned over backward in avoiding such actions, even as senior officials in the Pentagon and the White House were themselves engaged in such activity. Gerry, himself, even avoided public speeches during his tenure at ACDA, believing it best for the SALT negotiator to forgo the public forum.

In his dealings with Congress, however, he was open and accommodating, though somewhat hampered by the White House fixation on secrecy in the SALT negotiations. He believed that in spite of its inevitable problems, there would be value in having selected members of Congress as observers with the SALT Delegation, a practice that did not develop until later administrations came to office. In fact, his excellent relations with Congress and the high personal regard in which he was held there were probably his, and the Agency's, greatest protection against those in high places who might have wished to act against him and his agency.

One final relationship is worth noting, that with John McCloy and the General Advisory Committee (GAC) which McCloy chaired. Established by Congress in 1961, the GAC was mandated to advise the Director, the Secretary of State and the President on arms control matters. It was composed largely of senior establishment figures, and in reality many members of Congress in 1961 looked upon it as a watchdog of the newly created agency. McCloy had been its chairman from the beginning and had obtained a commitment from Mr. Nixon that, unlike the more routine review and observer role it had played in its first eight years, the Committee would be able to play an active role in policy advising.

Gerry worked closely with McCloy and with the Committee, which met frequently during the 1969–1972 period and became a strong supporting force for arms control within the government. As McCloy, himself, observed during a period in 1972, it was somewhat ironic that an establishment group set up in large part as a watchdog of ACDA had become a leading advocate of vigorous strategic arms control efforts. McCloy and the GAC joined with Gerry in the unsuccessful effort in 1972 to get Kissinger and the President to make a real effort, with serious proposals, to get a MIRV ban, as well as a "zero" ABM agreement.

The achievement of the 1972 ABM Treaty was for Gerry his greatest source of satisfaction during his tenure as Director of ACDA. The Interim Agreement on strategic offensive arms was to him a modest temporary restraint, for he was well aware at the time that the major failure of SALT I was the failure to ban MIRVs. But as he commented at the time, "a half-loaf is better than none." The Seabed Treaty, the Bacteriological Weapons Convention and the completion of arrangements for implementation of the Nonproliferation Treaty were significant, if lesser, accomplishments. And bringing the relatively new agency to a higher level of competence was a major positive development. However one allots credits for the achievements during these four years, Gerry's record as ACDA Director stands unmatched.

The White House-directed purge of the Agency that took place almost immediately after Gerry resigned in January of 1973 was a bitter reward for his services. For the Agency's personnel, the shocked reaction was perhaps reflected best in a literary offering penned at the time by Assistant Director Admiral Lee:

> The old crew at ACDA is canned.
> They misunderstood the command.
> They thought that their role
> was to plug arms control.
> At ACDA that subject is banned.

The record of achievement, however, remains. And the strengthening of the agency that Gerry accomplished helped sustain some professional competence in the organization's subsequent troubled times, and still stands as a model for those leaders who may seek to strengthen the government's pursuit of a more secure peace. As for Gerry Smith, after his hectic, stressful four years at ACDA, he left service as he had begun, a gentleman of quality.

Lawrence D. Weiler
Washington, D.C.
September 1988

The SALT Negotiations, 1969–1972

In a career of varied service both in government and in the private sector, the achievement for which Gerard Smith will almost certainly be most remembered is his role during the period 1969–72 as head of the U.S. delegation in the negotiations between the United States and the Soviet Union on strategic arms. These resulted in the Anti-Ballistic Missile Treaty and Interim Agreement on Strategic Offensive Forces, signed in Moscow on May 26, 1972 by Richard Nixon and Leonid Brezhnev. The Anti-Ballistic Missile Treaty is the most substantive and important arms control agreement ever reached by the two superpowers. It stands today, and is likely at least in modified form to go on for a long time, subject only to an unlikely success in the Strategic Defense Initiative during the 1990s. The Interim Agreement on Strategic Offensive Forces placed a cap on intercontinental ballistic missile and submarine launched ballistic missile launchers. It was criticized because it failed to limit warheads or throwweight (despite Gerry Smith's attempts to prevent deployment on both sides of multiple independently targeted reentry vehicles—MIRVs, whose existence make throwweight and total number of missile warheads important factors along with launcher numbers). But the Interim Agreement served a useful temporary purpose and was the first limit on strategic offensive forces that the U.S. and the Soviet Union had reached.

The story of the negotiations themselves is told in Gerry's book *Doubletalk* and in John Newhouse's book *Cold Dawn*. These go deeply into the substance of the negotiations. This brief account is intended more as a personal reminiscence of Gerry Smith and his role.

Someone must have been telling lies about me, because one fine day during the summer of 1969, Gerry Smith, whom I had met but didn't really know, called me in Pasadena to ask whether I would be willing to serve as a delegate to the negotiations that President Nixon had decided to initiate with the Soviets about limiting strategic arms. It was a second, and unexpected, chance; we are seldom presented with those, and I jumped at it (with some conditions). Following the Glassboro Summit in 1967, the Johnson Administration had hoped to initiate such talks and I, then Secretary of the Air Force, had come to be seen as the likely (part-time) Defense Department representative and number two delegate, with the Deputy Director of the Arms Control and Disarmament Agency, Adrian Fischer, leading the delegation. Soviet tanks in Prague had trashed that scenario, and in 1969 I had a full-time responsibility as President of the California Institute of Technology. But if Gerry would find it useful to have a part-time delegate, say three or four weeks every few months, with technical and strategic credentials, who would not be bound during the policy formulation by the position of any agency (though keeping a decent public silence on any disagreements), I was game. Thus began a fascinating association with the arms control process and with Gerry Smith, during which I was privileged to see his intellectual and human skills exercised on the most important task for Americans of our era—avoiding nuclear war while preserving U.S. interests.

Gerry assembled a varied and experienced team, who had among them distinguished pasts and promising futures. The formal number two was Phil Farley, who was also Gerry's deputy in the Arms Control and Disarmament Agency. Later on, Jeff Parsons an experienced and senior foreign service officer, took over the number two role. Paul Nitze, whose service to the United States in and out of government had lasted more than twenty-five years culminating in the office of Deputy Secretary of Defense, which he had left earlier that same year, and whose service in and out of government has continued, was Secretary of Defense Laird's representative. He brought great intellectual skill and force—and strong opinions. Tommy Thompson, former Ambassador to Moscow and by any standard one of the nation's two or three leading Kremlinologists, served until his fatal illness forced his retirement. General Royal Allison, representing the Joint Chiefs of Staff, had served in line and staff positions at the highest level in the Air Force. Raymond Garthoff, the delegation's executive secretary, was one of the most highly regarded of the State Department's younger experts on the Soviet Union. The delegation staff contained many outstanding individuals—among them John Rhinelander, David Aaron, Albert Carnesale, Sid Graybeal, Jim Woolsey, and Jack Shaw.

The Soviets had brought a delegation impressive in credentials and in skills. The chief of delegation, Nikolai Semenov had accompanied Marshal Konev's army into Germany at the end of World War II. He was an experienced diplomat. (Originally, it had been expected that Deputy Foreign Minister Kuznetsov, who had spent a year of post-graduate work at the Carnegie Institute of Technology, would be Gerry's opposite number. That he was not deprived us of what would have been an encouraging symmetry—kuznets is the Russian word for Smith. The formal number two was General Nikolai Ogarkov, later to occupy the most senior position in the Soviet military as Chief of the General Staff. Georgi Kornienko ran the Foreign Ministry's American desk. Academician Shchukin was familiar with ballistic missiles, nuclear weapons and electronics; he was a cultured intellectual who knew the national security business.

The U.S. delegation represented elements in the U.S. government, and the positions of those elements were reflected in the delegation. But being in Helsinki or Vienna (or later in Geneva) provided some separation and encouraged some individuality. Thus Gerry was faced with a complex set of bureaucratic and personal relations, which he managed skillfully. In dealing with the Soviets, his ability to speak passable Russian was a useful talent. And he parried Semonov's early polemics about Western imperialism by noting that his Irish ancestors had been stout resisters of British imperialism before emigrating to America.

By combining in his person the position of Director of the Arms Control and Disarmament Agency with that of Chief Delegate to the Strategic Arms Limitation Talks, Gerry intended to encourage a more coherent process of policy formulation. In fact this approach helped with some problems, but not with others. President Nixon and Assistant for National Security Affairs Kissinger clearly intended to have control of the negotiations. They did so through the interagency process in Washington, at which each agency tended to have its own agenda and positions. Some agencies, the Defense Department through the office of the Secretary of Defense and the Joint Chiefs of Staff, the State Department through the Foreign Service and the Arms Control and Disarmament Agency, had more than one. But Nixon and Kissinger also established a separate channel for negotiation through the Soviet Ambassador in Washington Dobrynin, and didn't always inform the delegation, the departments and agencies in Washington, or anyone else of what was going on in that channel.

In the meantime, Gerry had to handle social relations within the delegation, with the Soviets, and with the host governments—as well as the substantive negotiations. The Finns were delighted to have us, and with good reason; the SALT negotiation put Finland on the map as a recognized neutral and as a site for international negotiations, of which several have since taken place there. It may have made them feel safer from the Soviets. The Austrians displayed lavish hospitality—Vienna had, after all, hosted notable congresses before this. The first SALT session in Vienna was greeted by Kurt Waldheim, then Austrian Foreign Minister, but soon to be ousted in the 1970 election that brought Bruno Kreisky's socialist party to power. Later on, the Genevans were to prove more blase—they always had a dozen international conferences in session.

Along with the substance of the negotiations, Gerry had to look after the personal behavior of the members of the delegation. Among the more mentionable incidents were the Marines who rolled in the snow after saunas at the Hotel Kalastajatorppa in Helsinki and caught cold, and the member of the delegation staff on whose behalf Gerry had to intervene with the Viennese police after a brawl in the Prater, the Viennese amusement park. Trips north of the Arctic Circle, boat rides on the Danube, and sightseeing tours abounded; sometimes useful work was accomplished in these more informal settings. Gerry handled them very well, because he is a fine host and a pleasant guest, even though he is not a lover of purely social occasions. (His beloved late wife Bernice occasionally characterized Gerry as a bit stuffy.) After each formal meeting of the delegations, small groups from the two sides would go off with their opposite numbers to hold informal discussions and to have what the Soviets called "chai" (tea). The refreshments included alcoholic drinks as well. The formal public receptions involved champagne toasts. Photographs of receptions that marked the arrivals and the departures of the delegations from their respective capitals appeared occasionally in the press. Gerry, the most abstemious of men, was justifiably incensed when the Nixon-Kissinger team, which Watergate had not yet taught the need for alcoholic solace, sent him a stern admonition to cut out the alcohol—it was making a bad public impression.

After two and a half years of negotiations, agreement was reached. From Gerry Smith's standpoint, it was marked by end runs from the top and last minute decisions made without benefit of the knowledge and experience of the delegation. Some of those actions may well have been necessary in order to reach an agreement at all. Certainly some of the alleged defects in the SALT I agreements can be traced to those same actions. But none of the flaws, real or perceived, prevent the SALT I agreements from having been, and remaining, a success—the first post-World War II arms limitation agreement between the superpowers—nor do they lessen the credit due Gerry Smith for negotiating them (and Nixon and Kissinger for the policy decisions that went into them).

What can we say about the accomplishments of SALT I, and the lessons therefrom? First of all, the ABM Treaty, which is of unlimited duration, remains in force after sixteen years. As we continue in a time of technological change, it probably will need an update, but its principles stand. Their strength and its strength are shown by the failure of the challenge to the ABM Treaty even though that challenge was mounted by the most popular President in decades. Second, the structure of U.S. negotiating teams for later arms negotiations replicated that in the SALT I negotiations. This has been true in negotiations about mutual and balanced force reductions in Europe, intermediate range nuclear forces negotiations, anti-satellite systems negotiations, negotiations for limitations on conventional arms transfers, and others. The arrangements for agency representation; a backstopping committee in Washington constituted by representatives of the same agencies and dealing with daily instructions and reports from the delegation; an interagency group at the National Security Council level (which was called the Verification Panel in the case of SALT I), with a working group to do the staff work for that senior interagency group; congressional visits to the delegation as a way of assuring a knowledgeable congressional group when a treaty needs to be ratified by the Senate; frequent consultation with allies—the recent pattern was set at the SALT I negotiations.

It is almost certain that the President and his senior advisers will have to make decisions repeatedly in connection with any arms negotiations critical to U.S. security. But if negotiations are to succeed, the engaged delegations of the U.S. and Soviet Union will always have to explore the details, agree on all but a few of the issues, and highlight the remaining differences. Some small and some major issues will then be left over for resolution at the top by instruction to the delegations, or very often either through a separate channel or at a summit. But when the decisions are made by the President at the level of the principals (the Secretaries of State and Defense and the Assistant for National Security Affairs and perhaps one or two others), the SALT I history demonstrates the need for knowledgeability and care by those top officials. This should have been remembered at the Reykjavik Summit but was not.

Finally, in any negotiation dealing with U.S. national security, it is important that the negotiating team be headed by a person of the ability, experience, knowledge, dedication and integrity of Gerard Smith.

Harold Brown
Washington, D.C.
October 1988

Lt. J. G. GERARD C. SMITH, USNR, 1942

Gerard Smith with John Foster Dulles, 1958

President Nixon instructs his Chief Delegate to SALT I, 1969

Celebrating the signing of the SALT I accords, May 1972

Discussing nuclear non-proliferation goals with President Carter, 1978

Henry Owen and Robert Bowie with Gerard Smith, new recipient of the Presidential Medal of Freedom, 1980

The European Connection

Having known Gerard for over 30 years, what strikes me most looking back over that period is the consistency of his effort and method. His effort was to bring people, nations, and states, together; his method, to show that the problem that must be solved is essentially a common problem needing understanding of the other person's interest and point of view in order to arrive at solutions that can satisfy the parties involved.

Although I don't have a very good memory, the first meeting with Gerard that I remember very well took place during a cocktail party at his home. The circumstances were the following:

In late 1956, the Governments of France, Germany, Italy and the Benelux countries (the original six of the European Community), during negotiations that led to the setting up of the European Atomic Energy Community (EURATOM) and the European Economic Community (Common Market), asked three "Wise Men" (Louis Armand from France, Franz Etzel from Germany and Francesco Giordani from Italy) to report to them on the amount of atomic energy which could be produced in the near future in the six countries and the means to be deployed for that purpose.

I served as their principal staff-member. The oil crisis which resulted from the Suez affair had highlighted the danger of Europe's growing dependence on Middle Eastern oil.

Jean Monnet, whom I served at the time as Secretary General of his Action Committee for the United States of Europe, was very much interested in EURATOM, to a large extent his own brain-child. Atomic energy seemed to him a means towards solving Europe's energy problem, uniting six nations in developing this new source of energy and instituting close cooperation between the European Community, being formed, and the United States.

At the time, the United States was ahead of Europe in developing the peaceful uses of atomic energy; however, as a result of differences in energy prices, large scale construction of reactors was much more economical in Europe than in the States.

Furthermore, President Eisenhower's "atoms for peace" concept offered a strong political basis for cooperation. The three Wise Men decided to visit Washington in order to discuss possible forms of cooperation. Strong political support from the White House and the State Department were needed, as well as technical assistance from the United States' Atomic Energy Commission.

Monnet had discussed the matter already with the then Secretary of State, John Foster Dulles, and had laid the groundwork for the Wise Men's visit. Their task was to convince Washington of the seriousness of our undertaking (only a couple of years previously, the European Defense Community had ungloriously left the scene!) and to lay the basis for future far-reaching cooperation.

Since none of the three Wise Men had ever had direct dealings with the United States' administration, and the Atomic Energy Commission was pretty much a closed shop, the task was not an easy one.

Gerard was at that time Special Assistant to the Secretary of State for Atomic Energy Affairs. His understanding of the problem, his influence and his backing were essential to the success of the visit. It was at this cocktail party at Gerry's house that a real breakthrough towards full backing of the Atomic Energy Commission took place, some of us conspiring in between cocktails in a magnificent bathroom.

From that evening on, I have never had any doubt about the kind of aims Gerard pursued, nor about the efficacy of his method which includes, in my view, indispensable capacity for conspiracy.

For a whole range of reasons, EURATOM did not become as important as it had seemed likely after this cocktail party. However, if anyone is interested in studying an example of intense and trustful cooperation between the European Community and the United States, I can advise him or her to look at what happened during the years of EURATOM's conception and birth.

During the sixties, to a large extent as a result of the policies of General de Gaulle as President of France, relations between Europe and the United States deteriorated. Gerard then decided to start a monthly called

Interplay and asked me to help him in this enterprise on my side of the water.

Looking back at the history of European integration and the success which this process has in so many ways achieved, I am always struck by the number of defeats suffered along the way. To mention examples: the European Defence Community, the European Political Community, the United Kingdom's first attempt to join the Six, the effort, in the sixties, to set up a Multilateral Force, the Monetary Union plans of the early seventies, the failure of the so-called "snake" later on. As Jean Monnet used to point out, in the end it is strength of the process that overcomes the defeats. This is true, at least as long as there are men who continue to believe in and to stimulate the process. Gerard's monthly did not survive. However, the process overcame the transatlantic problems of the sixties.

For me *Interplay* did something more. Renewed cooperation with Gerard led him to invite Kathleen and myself to join Bernice and him to a voyage on the *Patra* from Naples through the Straits of Messina and along the Dalmatian coast to Venice. It became one of the most memorable holidays Kathleen and I ever spent. Back in our house in the Ardennes I wrote to Gerry and Bernice on the 1 September 1967: "I have only to close my eyes to see it all again: the first swim at Capri, Stromboli, a line of fire in the starry night, waking up before Panarea; crossing the Straits of Messina, the Dalmatian coast looming up in the light of early morning, Kôtor in its totally unexpected beauty at twilight, entering the Bay of Palaçe in the moonlight, smelling the warmth of land and pine trees; Vivaldi in the Rector's

Palace in Dubrovnik, looking down on the *Patra* from Marco Polo's house, seeing the moon set and the sun rise off Haver, a walk at Murter, a thousand islands and a thousand seas at our feet, the last sunny-hazy morning on the Adriatic and Venice in its splendour, opening up to *Patra*."

What I brought home from these summer weeks was not only the memory of all we had seen. At the end of that same letter I wrote: "We discussed together whether it was Augustine or Thomas who said: 'Esse ipsum bonum est'. We decided to ascribe it to Augustine. Now, despair, it seems to me, means being deeply shaken in the certainty of esse ipsum bonum est. Frankly, during the last six months, waiting for de Gaulle, seemed to contain more heartbreak than I could stand, and heartbreak sometimes is the beginning of despair. In a few days we will all be at our chores again. I do not think that there is much in the news to make us think that it will be easier than before. Probably, we have a few more nasty years in front of us. But the most precious thing of all that I take home from this cruise is that I have only to close my eyes to see again the incredible beauty of creation—then it will be a certainty to me again: esse ipsum bonum est."

Gerard and Bernice's hospitality did not limit itself to the *Patra*. Often during visits to Washington, I was their guest at dinners, mostly of a working nature, which happens to be the kind of dinner I enjoy most. On several occasions I stayed at their beautiful manor at Easton, Maryland.

During the most important period of Gerard's working life, the SALT years, I did not see much of him. However, one has only to read Gerard's *Doubletalk* to become well aware of the fact that he applied the same effort and method to negotiating that first and important arms limitation treaty just as he had done to keep Europe and the States on course through the Atlantic Alliance and through cooperation between the European Community and United States.

During the summer of 1972, David Rockefeller invited a group of Japanese, European and American friends to discuss his idea of setting up what was to become the Trilateral Commission. Our meeting encouraged him to go ahead and I was asked to look after the European side of the enterprise. Soon afterwards, retiring from his SALT assignment, Gerard agreed to preside over the North American side of the Triangle.

In the autumn of 1975 I took up residence in Florence as President of the European University Institute and, having to spend all my time on this new job, had to retire from the European presidency of the Trilateral Commission. However, from early 1973 to the end of 1975, I worked closely with Gerard during numerous Trilateral meetings, both in the United States, in Japan and in Europe. This is not the place to write the history of the Trilateral Commission, although I very much hope that some day, someone will do it. Let me, nevertheless, mention the aims of the Trilateral Commission as they were stated, early in 1973, in letters written to prospective members:

"The Trilateral Commission should be an independent body, constituted by citizens of the European Community, the United States and Japan, pursuing the following aims:

a) to propose policies which the Member States of the European Community and the European Community itself, the United States and Japan could follow in

1. their economic relations;
2. their political and defence relations;
3. their economic and political relations with developing countries;
4. their relations with the countries of Eastern Europe and with China;

b) to gain acceptance of these policies."

Now that this Commission has brilliantly come of age and is making a regular contribution to the fulfillment of these aims, it is not easy to imagine how difficult the beginning was.

Only members of the Executive Committee took part at the first formal meeting in October 1973 in Tokyo, and in June 1974 in Brussels, while France was represented only by an observer. Between these two meetings, the price of oil shot up fourfold. The ensuing economic dislocations led to a begger-your-neighbour, 'save yourself first' mentality, far removed from Trilateral ideals. There were moments when it seemed impossible to continue the Commission's work. Also in those early years there was nothing of the free and uninhibited discussions, in any case not from the Japanese side, which now have become routine.

Henry Kissinger's "Year of Europe" declaration, however well intentioned it may have been, did not help matters, because, especially in hesitant France, the Trilateral Commission was unjustly seen as another instrument of the United States' policy, especially in the energy field, where tensions were real. However, the momentum created by David Rockefeller and maintained by Gerard and Zbig Brzezinski as director of the Commission carried us through. Once again, the process proved itself strong enough to overcome the difficulties which it encountered.

When the full Trilateral met at the end of 1974, and a decision for a May 1975 meeting in Kyoto was taken, it was clear that the Commission was there to stay. Of the many in-between meetings of the Chairmen and Zbig (and several others, including the initiator of the enterprise, David Rockefeller) I vividly remember one in particular in the autumn of 1974 which had the task of finishing preparations for the Washington full meeting a few months later. It took place in Gerard's lovely and stately home at the Eastern Shore of Maryland. Even if the meeting had lasted as many months as it did days, I would not have become homesick, so much this place looks and feels like one of the country houses the patricians of Amsterdam built themselves in the 18th century, along the the calmly winding Vecht River!

From Florence, and after my return to Brussels, I kept up with Gerard's thoughts and actions through the articles he wrote, mostly on arms control and often together with others like McNamara and George Kennan.

Coming back to Washington in 1988 after a long absence, I enjoyed once again, in his new home, his old hospitality. The discussion round the dinner table proved that he had neither abandoned his effort nor changed his method. Neither disappointments, part of every life devoted to action, nor the personal loss he has suffered have encroached on the conviction he exposed, cruising on board the *Patra:* esse ipsum bonum est.

Max Kohnstamm
Brussels, Belgium
September 1988

Nuclear Nonproliferation, 1954–57 and 1977–80

Over a span of four decades, Gerard Smith has actively engaged in efforts to constrain the spread of nuclear weapons. Threatening both national and international security, nuclear proliferation, should it occur, would undermine the fragile barriers of nuclear self-restraint in effect since August 1945 and raise the odds that nuclear weapons would once again be used.

In 1989, he is still active as U.S. Chairman of a Joint U.S.-European effort to energize new nonproliferation initiatives. But in the early 1950s at the U.S. Atomic Energy Commission, and from 1954-1957 as Special Assistant for Atomic Energy Matters in the State Department, he was already engaged in efforts to devise and bring into existence international barriers to nuclear proliferation. He saw the importance of the United States acting responsibly in its own nuclear activities, to avoid facilitating nuclear weapons capabilities. He saw also, and this has been less common, the limits to what the United States could achieve through unilateral denial and compulsion. In the long run, nonproliferation can only be viable on a basis of voluntary cooperation, under circumstances where participating non-nuclear nations find their own security served.

Special Assistant (Atomic Energy) to the Secretary of State (S/AE, 1954–57)

Nuclear nonproliferation moved gradually onto the international agenda in the aftermath of President Eisenhower's "Atoms for Peace" speech to the United Nations General Assembly in December 1953. Initially after World War II the United States had sought to maintain an atomic monopoly on civilian as well as military applications, by means of secrecy and autarchy. But Soviet and British nuclear tests in 1949 and 1952 respectively showed the futility of such a policy of denial. The speech was a dramatically effective reversal of U.S. monopolistic policy, including a proposal that the nuclear powers initiate a program of drawing off nuclear materials from their weapons arsenals and making them available for peaceful uses to other interested nations through a new international agency.

The speech unfortunately greatly overstated the effect that such a program could have on the rapidly expanding nuclear weapons stockpiles of the Soviet Union and the United States. It even more grossly oversold the prospective benefits of nuclear power, especially for the developing nations, who were told that a "special purpose would be to provide abundant electrical energy in the power-starved areas of the world." Only incidental reference was made to the need for caution. Gerry would much later recount the reception the speech had at the time in some sophisticated quarters:

> . . . while working at the Atomic Energy Commission in 1953, I happened to have been directed to read President Eisenhower's U.N. address to those members who had no advance notice. One knowledgeable Commissioner remarked that it was a "thoroughly dishonest proposal." Later when Molotov protested to a dubious John Foster Dulles that the proposal would result in the spread worldwide of stockpiles of weapon grade material, I had to explain to Dulles that Molotov had been better informed technically than he. Subsequently, the Soviets asked how we proposed to stop this spread. The best we could reply was "Ways could be found." We are still searching for those ways.[1]

Carrying out the new U.S. policy required major revisions in applicable U.S. legislation, to permit U.S. cooperation with other countries not only through the new International Atomic Energy Agency (IAEA) but also bilaterally on the peaceful uses of atomic energy. Allies would be given a limited participation in nuclear operations (such as the NATO two-key system under which in actual hostilities U.S. warheads would be released from U.S. custody for allied delivery). New tasks fell to the Department of State, and Dulles asked Gerry to become his personal adviser, heading a small office. He and his staff worked closely with the Atomic Energy

---

[1]*Gerard Smith on Arms Control* (Kenneth Thompson, ed.), University Press of America, 1987, p. 117

Commission on agreements for peaceful atomic energy cooperation, and with the Department of Defense on arrangements for military cooperation. They also represented the State Department on the Joint Atomic Energy Intelligence Committee chaired by CIA, and coordinated arrangements (mostly secret) with friendly countries for installing monitoring equipment directed at Soviet tests. With the State Department's Bureau of European Affairs and the Policy Planning Staff, Gerry took a particular interest in nuclear weapons policy and related NATO doctrine and deployments.

Once the Atomic Energy Act of 1954 was passed, the United States took the initiative to enter into agreements for peaceful atomic energy cooperation with dozens of countries. These agreements barred use of exported information, materials, or equipment for any military applications, and provided for inspection by U.S. teams—a partial answer to Molotov's question. But it was clear that the United States could not alone dictate and police a global nonproliferation regime. Gerry led the interagency group charged with fleshing out the President's proposal of a new IAEA, and served as deputy to Ambassador James Wadsworth in the gradually broadened negotiating groups (with the Soviet Union particpating from the outset, albeit somewhat passively and suspiciously) culminating in establishment of the Agency with U.N. blessing in 1957. The safeguards functions of the IAEA, to detect diversion to weapons purposes, was soundly if generally conceived and articulated in the Agency statute. Such multilateral responsibility for monitoring programs and deterring proliferation was, over the long term, more firmly based and more effective than unilateral U.S. policing. As the IAEA developed safeguards staffing and competence, the United States transferred the safeguarding of its bilateral arrangements to the IAEA staff.

The Atoms for Peace program was highly promotional. The United States pressured countries in Latin America and elsewhere to sign bilateral agreements even when lacking any indigenous atomic energy competence. Gerry was uncomfortable but could only occasionally apply the brakes. One small victory came when the idea surfaced in the White House that the program could be accelerated if we simply *gave away* small research reactors. Gerry argued vigorously that such visible donations would in many cases become highly visible white elephants, to the embarrassment rather than the credit of the United States. The U.S. offer was then changed to one on a matching funds basis—sufficient to eliminate most of the unqualified recipients (though white elephant research reactors were not an extinct species). Not only was the U.S. Atoms for Peace program more reasonable; a gain for nonproliferation resulted, since such research reactors were fueled with small but worrisome amounts of weapons-grade highly enriched uranium, a source of concern even if subject to inspection at intervals.

In alliance with Robert Bowie, the head of the Policy Planning Staff, Gerry worked diligently to bring Dulles to accept that massive retaliation was an unrealistic strategy in the face of a growing Soviet retaliatory capability in the mid-1950s. He also worked persistently to get the case for a nuclear test ban examined objectively. After the dust settled from the 1956 presidential election, in which Stevenson espoused a test ban and Eisenhower rejected the idea, it became respectable to urge that a test ban be seriously considered—among other reasons, as a nonproliferation measure. Gerry had established a good rapport with Dulles, on many common grounds including legal training—and, in informal sessions, a liking for somewhat competitively quoting Biblical passages more or less relevant to the issue at hand.

When the White House based Stassen disarmament effort collapsed and the disarmament coordination responsibility reverted to the State Department, it came to rest soon with S/AE. When Bowie returned to private life in the fall of 1957, Gerry succeeded him as Assistant Secretary for Policy Planning. He continued his interest in nuclear and arms control matters, but operational aspects of nonproliferation devolved on others in State and associated agencies.

Special Presidential Representative for Nonproliferation Matters 1977-80

As Director of the Arms Control and Disarmament Agency from 1969 to the end of 1972, Gerry continued active support of U.S. nonproliferation efforts. ACDA had the leading role within the executive branch, in coordination with the Atomic Energy Commission and the State Department. But the demands of the protracted SALT negotiations and other responsibilities of his position left him little time for nonproliferation. With the Carter administration, and his appointment in mid-1977 as Ambassador-at-Large and President's Special Representative, nonproliferation matters became his full-time concern.

In the mid 1970s, following the 1973 "oil shock" and the prospect of accelerated nuclear power programs around the world, nonproliferation concerns heightened in the United States. One focus in the public and in Congress was the prospect of the emergence of a "plutonium economy," with dire impact on the new environment as well as on hopes for stopping the spread of nuclear weapons. In the final year of the Ford Administration, President Ford and Secretary of State Kissinger announced stepped-up non-proliferation efforts, both national and international, and a restrictive U.S. approach either to plutonium separation in the United States or to export of the technology. President Carter carried these approaches further, in one of the first highly publicized initiatives of his administration. In April 1977, he called for revision of U.S. legislation (what became the Nuclear Nonproliferation Act of 1977); it included such provisions as U.S. veto rights over foreign reprocessing or other disposition of spent fuel supplied by the United States or discharged from U.S.-supplied reactors. Renegotiation of existing agreements and contracts was also required, with tightening of export license criteria in a way seen by U.S. partners as unilateral amendment of agreements. The U.S. explanation and justification of these measures carried a strong flavor of rejection of nuclear power in general, and of the breeder reactor or other use of plutonium fuel in particular.

With the exception of Canada and Australia, where there were similar currents of public and parliamentary opinion, and where nuclear power was not a major energy source, world reactions were strong and hostile, among U.S. allies as well as most other countries where nuclear power was of current or prospective interest. They resented especially the unilateralism of the U.S. approach, the call for renegotiation of existing agreements, the preaching and at times hectoring tone, and the pre-judgment of complex and difficult energy options and nuclear power technical choices. It was immediately clear that the United States had to reshape its presentation, and perhaps some of its premises.

Secretary of State Vance and his deputy Warren Christopher were old Washington hands to whom Gerard Smith was well known. In addition, Gerry had known and worked with Jimmy Carter and Zbigniew Brzezinski, as U.S. Chairman of the Trilateral Commission from 1973 on. He had recruited Carter as a member; Brzezinski was Executive Director of the U.S. Commission staff. When the counter-productive effects of the nonproliferation crusade became evident, it was natural for them to call on Gerry to help mend fences and put the U.S. program back on the right track.

Gerry was indeed an ideal choice. He was known and respected in Congress, not least for his nonproliferation credentials going back to the 1950s. He was equally well known and respected in Japan and Western Europe through his private and governmental activities in many fields going back twenty years before his Trilateral Commission role. His contacts with the Soviet Union also long preceded his SALT negotiator role, in particular in the IAEA negotiations from 1954-57. He was a strong supporter of the IAEA from that time on, and in accepting the new appointment, he stipulated that he also be named U.S. Member of the IAEA Board of Governors.

In starting work, he typically rejected advice from bureaucratic warriors that he function as "nonproliferation czar" with offices in the White House. In line which his longstanding respect for the role of the Secretary of State as coordinator of foreign policy and operations, he insisted on being located in State, reporting directly to the Secretary and his deputy, both of whom were strong backers for Gerry's concept of his role. He was not enthusiastic about the extreme provisions of the emerging Nuclear Nonproliferation Act. Since drafting was well advanced and the Administration was quite firmly committed, he left the completion of the task to others. In general, he left other interested offices in State, the Department of Energy, and the White House to go about their business, and provided himself with an exceedingly small staff to keep in touch with the others and support him.

He took as his principal missions to restore cooperative and trusting relations with allies, and where possible with the so-called "threshold states" who were interested in but not committed to the nuclear weapons option. He did take over responsibility for one incipient activity to this end. Even before his entry on the scene, Carter administration officials, in a damage-limiting and time-buying effort to put a more cooperative face on the Carter nonproliferation initiative, had devised the concept of an "International Fuel Cycle Evaluation" (INFCE). INFCE was to be a joint technical examination by experts from all interested nations of some of the main Ford/Carter propositions—the risks attendant on the "plutonium economy," the adequacy of nuclear fuel supplies without resource to the plutonium breeder, the preference for the "once through" fuel cycle as more "proliferation resistant" than recycling plutonium, and the desirability of proscribing "sensitive nuclear fuel cycle facilities" (reprocessing, enrichment, heavy water production, the breeder) for all countries except perhaps a very few with "good nonproliferation credentials" and well-warranted nuclear power needs. Gerry leaned over backward to show the United States as coming to listen to others and to understand their needs, concerns, preferred courses of action rather than to make a hard sell for U.S. proposals. He insisted that no new international staff was called for, and put the IAEA in its proper position as providing the meeting site and staff support for the venture.

INFCE turned out, not coincidentally, to be a contribution to the education of the United States, and not only to the rest of the world. From the perspective of the Carter administration, it was in the words of a *Washington Post* editorial at its conclusion in early 1980, "a modest success." Other countries remained firm on their position that they would decide their own nuclear power needs and technical approaches, but with keener appreciation of the risks accompanying sensitive stages of the nuclear fuel cycle where nuclear weapons-useable materials are produced or present. And there was evident, in INFCE and in side dealings, a greater readiness to examine ways of buttressing nuclear nonproliferation measures with additional safeguards against such risks.

INFCE was not, however, Gerry Smith's main preoccupation over these years. His basic premises, which underlay his activities of choice, were simple. Proliferation of nuclear weapons is a real threat. Its motive is insecurity. Dealing with it requires dealing with the roots of insecurity. Safeguards and some discriminating restraints on national nuclear programs and international nuclear trade can help stave off proliferation. But to be effective they must be based on assent and cooperation, not denial. And even cooperative technical or safeguarding nonproliferation measures could not be decisive. Many nations have the potential capability to produce nuclear weapons if they judge them necessary for their security. Thus, the key is motivation, and to affect that desire the international community had to find ways to make the threshold states less insecure. For that, there are no quick, tidy or easy ways.

Gerry, thus made it his special task to tackle the politics of non-proliferation. There were two principal groups to deal with: our major allies—Japan and the members of NATO; and the threshold nations—India, Pakistan, Brazil, Argentina, South Africa, Israel.

Reknitting relations on nonproliferation and nuclear power with our major allies bore on nonproliferation in two ways. Their cooperation in maintaining and strengthening the nonproliferation regime was of course crucial, since they were potential or actual suppliers of technology, equipment, fuel, and services. Even before that, as Gerry had stressed for many years, they were themselves among the successes of nonproliferation policy—that Japan and Germany were not nuclear powers was made possible by our security guarantees under our alliances. And it must be borne in mind that, in ratifying the Nonproliferation Treaty, those two countries made formal stipulations that their commitments were conditioned on their alliance with the United States, and could be reexamined if circumstances changed.

One of Gerry's first undertakings was to defuse the U.S.-Japanese dispute over the pilot Japanese reprocessing facility at Tokai Mura. The pending U.S. legislation would make it vulnerable to U.S. actions. The energy-poor Japanese, with their bent for long-range planning, saw the facility as essential to their very substantial nuclear power program. Gerry put together and led a U.S. interagency group that in Tokyo and Washington negotiated an agreement under which the Japanese would be able to go ahead, though dependent on U.S. good faith in view of its assertion of veto rights under the NNPA. The *quid pro quo* for the U.S. was essentially Japanese commitment to participate actively, bilaterally and in INFCE, in programs to reinforce the nonproliferation regime. In working out this resolution, Gerry made good use of his personal standing and close relations with senior Japanese officials, and also his access to Carter and Brzezinski, to get the substance of the agreement endorsed and published quickly so that it would not become lost in the feuds on nuclear power issues in Washington.

As for Europe, there were already many in the administration and Congress who were alert to the need—for nonproliferation itself—to find common ground with European allies with their role as alternate nuclear suppliers. It was thus a matter for Gerry of lending support to successful efforts to get special legislative provisions easing the application of denial clauses to Europe. Then it was a matter of putting back on a basis of mutual confidence the continuing dialogue with the French, West Germans, British, and others regarding INFCE, problems of supply to the threshold states, and other issues continually emerging or recurring. The French and Germans were already reexamining their export policies on sensitive fuel cycle facilities, and moving to recognize the need for restraint and selectivity.

Among the threshold states, there were several with whom the United States had close security ties, or a continuing dialogue, and at times action on problems of regional security, which provided the broader context for containing proliferation. Israel and Pakistan are obvious examples. There was also an involvement with South Africa in efforts to ease and resolve the tensions and even conflicts in southern Africa. In the case of South Africa, Gerry was called on to visit that country and meet with senior officials, after the apparent South African preparations for a nuclear test gave greater urgency and relevance to clarifying South African nuclear goals, preferably by signature of the Nonproliferation Treaty. He also made several unpublicized visits to Brazil and Argentina to add his arguments to the U.S. case that bringing into force the Treaty of Tlatelolco (the Latin American Nuclear Weapon Free Zone, with IAEA inspection of all nuclear activities of the member states) would be in the security interests of the region and of the individual countries, as well as facilitate international trade and cooperation of importance for their nuclear programs. Decisive results cannot be claimed for those interventions, but they were a factor in keeping policies fluid and deferring any commitment by these states to moving definitively to the nuclear weapons options. He also played an important role, especially with the Congress, in finding a way to prevent punitive application to India of the U.S. denial rights asserted under the NNPA. The effect of such actions would have been awkward—but no more—for India's nuclear program. The cost would have been loss of such influence as the United States still possessed over future Indian nuclear decisions.

A period of damage limitation and course corrections—Gerry Smith's role in 1977-80—is not a dramatic one. Nor can it be a public one if carried on in part to correct policies of the President for whom one works. Thus he could not state publicly what he was trying to do until after the period was over—in the article he wrote jointly with colleagues George Rathjens ("Reassessing Nuclear Nonproliferation Policy," *Foreign Affairs,* Vol. 59, No. 4, Spring 1981, pp. 875 ff.).

But nonproliferation is perhaps in essence a long-term damage-limiting effort. Long-term, and also against the odds in a world with all too many insecure and hostile industrializing neighbor nations, while the two superpowers are still not so much reducing their massive nuclear arsenals as slowly and warily restructuring them. The patient task is to hold together an uneasy concert of nuclear powers and non-nuclear weapons states, and of energy-rich and energy-poor economies. To hold together uneasy constituencies, in the United States as well as in other countries, simple false linkages must be rejected—"Nuclear power brings nuclear weapons;" "forswearing nuclear weapons means second-class political status;" "nuclear weapons are the direct way to national security."

By mid-1980, it was apparent on the domestic scene and at the final session of INFCE that the United States remained part of the central consensus on nonproliferation that held these diverse international pulls in balance. The timing was fortunate. The second five-year review conference in 1980 was to examine the working of the Nonproliferation Treaty of 1970, the cornerstone of the regime. The conference was a fractious one. The collapse of SALT II and increasing U.S.-Soviet tensions raised deep concerns about the superpower side of the NPT bargain. Continuing energy supply problems caused many countries to remain determined not to give up the nuclear option. In this atmosphere, the review conference made no new nonproliferation progress. Nevertheless, a clear if unexpressed consensus was evident that nonproliferation remained important and that the NPT and the IAEA safeguards system were sound and should be maintained. (A similar and somewhat more explicit outcome was to emerge from the third review conference in 1985, under somewhat more favorable circumstances.)

In the United States, nonproliferation did not become a party issue in the 1980 elections or in Congress. When the new Reagan administration assumed office, it appointed a knowledgeable and committed career official as Ambassador-at-Large and U.S. representative to the IAEA. Even when U.S.-Soviet tensions and recriminations were at their height, and arms control negotiations were suspended or adversary, nonproliferation consultations and cooperation went ahead between the two rivals. The policies on nuclear power and nuclear nonproliferation and their interaction which were formally issued by President Reagan and pursued by his representatives did not differ markedly from those set forth in the Smith-Rathjens article mentioned above.

Even though mostly behind the scenes, Gerry Smith's role in the continuity of U.S. nonproliferation policy was well understood internationally—at the IAEA, among U.S. allies, in the Soviet Union, and among threshold states. He was of course not the only actor. But it is not too much to claim that his role was unique in restoring confidence in the U.S. course on nonproliferation and related arms control and nuclear power programs. The roots of this benign influence are simply: the consistency of his long career of related public service, in many agencies and under many Presidents and Secretaries of State; and his transparent personal integrity.

Philip J. Farley
Stanford, California
October 1988

The Arms Control Association and the Reagan Years

In 1980, prospects for arms control had never looked bleaker. The new President was on record against the principal arms control agreements and proposals of all administrations since Truman and he was calling for America to seek decisive military superiority. When most men who had already done far more than their share in fighting the good fight for arms control would have turned to cultivating their own garden on the banks of the Chesapeake Bay, Gerard Smith chose not to leave the field of battle but to hold aloft the standard of sensible arms control through the confused, and at times frightening, eight years of the Reagan Presidency. As Chairman of the Arms Control Association and in a number of related activities, he labored hard, first to save the existing arms control regime, which he had played such an important role in establishing, and then to support and advocate new initiatives.

At the ACA Annual Meeting on December 17, 1980, in one of his first acts on leaving the Carter Administration, Gerry delivered an address entitled, "There is No Other Way," in which he anticipated the struggle to come by predicting that the new Administration would try to modify, if not terminate, the ABM Treaty. He warned that, "This threat to the ABM Treaty is the greatest danger to the present structure of international arms control." Despite the seemingly dismal prospects for arms control, he advised his audience that there was still life in the arms control process and cautioned that, "if humans are to survive much longer, there is no other way than that of international control of nuclear weapons." Shortly after delivering this address, Gerry joined the Board of ACA and later in 1981 became Chairman of the Board, replacing the ailing William Foster, who had founded the Association in 1971 after serving as the first Director of ACDA under Presidents John Kennedy and Lyndon Johnson.

True to his campaign promises, President Reagan demonstrated his intention to expand U.S. military might and his disinterest in constraining arms. Nevertheless, reacting to growing domestic discontent with his military policies, and to pressure from the NATO allies, he resumed negotiations with the Soviet Union in the fall of 1981 on Intermediate Nuclear Forces (INF). In the face of mounting criticism, the United States finally also resumed negotiations on strategic arms in mid-1982 which were designated the Strategic Arms Reduction Talks (START) to distinguish them from the "fatally flawed" SALT discussions. The initial positions of the two sides were far apart, and little progress was made in the initial phase that lasted through the end of 1983.

The prospects of the arms control regime were further

dimmed in March 1983 when President Reagan gave his famous Star Wars speech, calling on the scientific community to create an impervious strategic defense that would make nuclear weapons "impotent and obsolete." This radical proposal not only flew in the face of the existing technical consensus that it would be impossible to achieve such a defense against a sophisticated and determined adversary but it also called for a fundamental change in strategic doctrine. By rejecting as unacceptable mutual deterrence between the two superpowers and their allies, this proposal challenged the foundation on which both national security planning and strategic arms control had been built.

The most immediate target of Star Wars appeared to be the ABM Treaty which had been designed specifically to prevent provocative attempts, however technically misguided, to develop, test and deploy a nationwide missile defense system. The other victim of the new strategy appeared to be any prospects for a START Treaty since the Soviet Union made it clear that it would not reduce its strategic offensive forces significantly in the face of a massive ABM deployment by its principal adversary. Moreover, Gerry often wondered aloud whether the U.S. Joint Chiefs of Staff would themselves ever agree to a START agreement severely limiting U.S. offensive forces if the Soviet Union could deploy massive missile defenses.

At the end of 1983, the Soviet Union walked out of the INF and START negotiations after the United States began the deployment of Pershing II and ground-launched cruise missiles in Europe. Through 1984, the impasse continued as the Strategic Defense Initiative

(SDI) took shape in response to the President's challenge.

When I arrived on the scene as the Association's new Executive Director in January 1985, Gerry had been Chairman of the Board for over three years. Working with Association President Herbert (Pete) Scoville, Gerry had taken the lead in a damage limiting campaign to protect the existing arms control regime from attack by the Reagan Administration. Publicly and privately he used his considerable influence to defend both the ABM Treaty, which was coming under increased attack as the SDI program gathered momentum, and the quantitative limits established in the SALT II Treaty, which the Reagan Administration formally announced it would never ratify.

When confrontation over the future of the ABM Treaty grew more intense after the Star Wars speech, Gerry did not take an academic or elder statesman approach but joined the fray by accepting the Chairmanship of the National Campaign to Save the ABM Treaty. This umbrella organization brought together representatives of ACA and other public interest groups, including: the Union of Concerned Scientists, the Federation of American Scientists, the Council for a Livable World, Common Cause, and the League of Women Voters.

Any reservation I might have had about working with the Association after my first view of the cramped quarters its small staff then occupied at the Carnegie Endowment for International Peace were quickly overcome by the opportunity to work with my old friends, Gerry Smith and President Pete Scoville. In discussing with Gerry whether or not to take on the directorship of the

Association, my thoughts went back to my first conversation with him sixteen years earlier and our subsequent activities. In early 1969 when I was on Henry Kissinger's NSC staff as one of the few holdovers from the Kennedy/Johnson period, I received a totally unexpected phone call from Gerry, whom I had met at meetings but did not really know. Coming immediately to the point in his inimitable style, Gerry asked if I "would consider" leaving the White House to be his Assistant Director at ACDA responsible for preparations for the proposed SALT negotiations. As a man whose decisiveness I later came to admire, Gerry must have been favorably impressed by my ability to accept his offer on the spot. Later he may have realized how welcome the opportunity was to escape from an NSC staff which at that time made even the floor of the jungle seem "a kinder and gentler" place.

Gerry's crucial leadership role as the negotiator of the ABM Treaty and the SALT I Interim Agreement and Director of ACDA are discussed elsewhere. I will only remark that my close working relationship with him during that period was one of those experiences that make government work so rewarding. Here was a man who combined determination to produce results with unimpeachable integrity, an open mind, and an insatiable desire to learn more about the problems he was trying to solve.

We worked together again in the Carter Administration where he served as Ambassador-at-large for Nonproliferation Affairs. As Deputy Director of ACDA, I was preoccupied with SALT II and other U.S./Soviet negotiations and found it most reassuring to know that the little appreciated, but critically important diplomatic negotiations to restrain proliferation were in Gerry's able hands. I grew to appreciate even more Gerry's unflagging determination to find solutions to dilemmas that seemed to involve insurmountable political and bureaucratic problems.

With these personal reflections in mind, I had no difficulty in making the decision to accept the directorship of the Association at a time when the very survival of the arms control process seemed in doubt. None of us knew then that Pete would shortly suffer a recurrence of cancer and after a heroic fight would succumb that summer. Subsequently, in the dual capacity of President and Executive Director of the Association my relationship with Gerry became even closer.

When I started at the Association in January 1985, I was greeted by a five-year plan that Gerry and Pete had been developing to guide the activities of the Association. I was very pleased to have in hand what seemed to be a realistic plan of action for the second half of the '80s. At the beginning of 1989, I believe their plan still provides sound guidance on how the Association should enter the decade of the '90s.

Highest priority was to be given to maintenance of the ABM Treaty and containment of the SDI program so that it would not destroy the treaty. Second priority was given to controlling strategic arms, including continuation of the SALT II limits and the achievement of reductions without loss of crisis stability. Third priority was given to developing public understanding of verification and compliance issues so that this would not become a barrier to progress in arms control. Fourth priority was assigned to nuclear nonproliferation, including both more aggressive implementation of existing measures and new reinforcing initiatives such as the Comprehensive Test Ban and the cut-off of production of fissionable material for weapons. Finally, fifth priority was given to developing the concept of no-first-use of nuclear weapons, a proposal that Gerry had joined McGeorge Bundy, George Kennon, and Robert McNamara in advocating in a now mini-classic *Foreign Affairs* article in 1982.

Prospects for progress also appeared a bit brighter when later in the month (January 1985) the Soviet Union agreed, after a year away from the negotiating table, to resume negotiations on the START and INF Treaties under the umbrella of new Nuclear and Space Arms Talks that would also include space weapons. Since the Soviet walkout in late 1983 had failed to stop the U.S. deployment of INF missiles, many believed that the resumption simply reflected Soviet recognition that its boycott had become tactically counterproductive. A much more important development that was not recognized at the time was the emergence in March 1985 of Mikhail Gorbachev as the new Soviet General Secretary.

Despite erratic swings and moments when disaster seemed imminent, U.S./Soviet relations began a slow climb from the depths reached during the first Reagan term. At their Geneva summit in November 1985, Reagan and Gorbachev agreed to the goal of a 50 percent reduction in their strategic arsenals. Two months later, Gorbachev proposed the phased transition to a nuclear free world by the year 2000, establishing an idealistic vision to match Reagan's vision of a world free from the threat of nuclear weapons under an impervious Star Wars shield.

The otherwise improved prospects for progress were dimmed, however, by the efforts of the Reagan Administration to re-interpret the ABM Treaty to facilitate the SDI program. In October 1985 the United States endorsed as "legally correct" the so-called "broad" interpretation of the ABM Treaty. The implementation of this decision would have in Gerry's words made a "dead letter" of the treaty. Thus, as the pace of negotiations on a START Treaty began to quicken, the fundamental difference on the future of the ABM Treaty intensified. This impasse remained through the end of the Reagan Administration and prevented a START agreement that otherwise appeared within reach.

As the negotiator of the ABM Treaty, Gerry took particular umbrage at the "broad" reinterpretation, which he saw as a meritless and cynical effort to destroy the treaty. He took the lead in rallying the support of former government officials, including six Secretaries of Defense and key Senators in support of the Treaty. His prestige helped the Association in its continuing educational campaign that certainly played a significant role in saving the treaty at a series of critical junctures. If the ABM Treaty had not been saved, there would not have been any of the subsequent progress on arms control that we have witnessed and U.S./Soviet relations would probably not have evolved in anywhere near as favorable a fashion as they have.

In May of 1986, the arms control process suffered another blow when President Reagan announced that, as a consequence of alleged Soviet violations of the unratified SALT II agreement, the United States would no longer be bound by its "political commitment" to the limits of SALT II. Gerry again lent his prestige to the Association's efforts to assess objectively the ambiguous compliance issues which the administration exploited to terminate the SALT II Treaty. The unpopular task of questioning one's own government's complex charges against the "evil empire" played a critical role in legitimizing objective discussion of the administration's mounting attack on the Soviet compliance record. This was an important step since the attack ultimately appeared directed at the ABM Treaty and the arms control process itself.

Then the world was taken by surprise in October 1986 when Reagan and Gorbachev met on very short notice in Reykjavik, Iceland. The ensuing exchange covered a remarkable range of radical proposals, including a total ban on all ballistic missiles and even the elimination of all nuclear weapons. Although the meeting eventually collapsed in disarray and acrimony over the old issue of strategic defenses and the ABM Treaty, some progress was made on establishing the framework for the START and INF Treaties.

The Iran/Contra scandal that followed hard on the heels on the Reykjavik summit threatened to paralyze the Reagan administration and to destroy the progress that had been achieved toward new arms control agreements. In an apparent effort to regain the initiative, the Reagan administration in November 1986 deliberately exceeded the limits of the repudiated SALT II Treaty. At the beginning of 1987, the administration apparently even seriously contemplated implementing the "broad" interpretation of the ABM Treaty as part of a program for "early deployment of a nationwide ABM system," in violation of the ABM Treaty. A precipitous decision may have been forestalled when Senator Sam Nunn formally advised that such a unilateral reinterpretation would create "a constitutional confrontation of profound dimensions." Gerry, who as the chief negotiator of the ABM Treaty had testified before the Senate in its ratification proceedings, played a major role in the Association's efforts to assist the Senate in this epic struggle. The issue evolved into a protracted debate between Senator Nunn and the State Department's Legal Counsel, Abram Sofaer. In this debate, Gerry and the Association contributed significantly to the general effort to hold the line against the administration's persistent efforts to reinterpret the treaty.

The downward drift of events was dramatically reversed in February 1987 when Mikhail Gorbachev agreed to delink negotiation of an INF Treaty from the overall strategic negotiations and essentially agreed to accept the U.S. "zero option" position. Reagan accepted "yes" for an answer, and the balance of 1987 was taken up with negotiation of the many details of the complex treaty. With a major effort, the INF Treaty was signed at the Washington Summit in December 1987. While limited in coverage, the treaty abolished a particularly threatening class of weapons and created an important precedent for intrusive on-site inspection as a verification measure. When the treaty was approved by the Senate by a vote of 93-5, bipartisan support was reestablished for the arms control process. Gerry and the Association supported the INF Treaty on its own merits and as an important step in improved U.S./Soviet relations.

The Washington Summit also produced the framework for a START Treaty and predictions were made that the treaty would be ready for signature before summer. However, the underlying impasse on strategic defenses persisted and little progress was made on the remaining issues of START, such as the treatment of sea-launched cruise missiles. The Moscow Summit at the end of May 1988, while reflecting further improvement in U.S./Soviet relations, failed to produce a START agreement as many had hoped.

There was no significant progress on arms control in the final six months of the Reagan Administration. There was in fact a ripple of discord in the fall at the Five Year ABM Treaty Review Conference when the two sides exchanged charges of treaty violations, and the United States threatened to charge the Soviet Union with a "material breach" of the treaty for having deployed the Krasnoyarsk radar that was contrary to treaty provisions. But it was soon clear that Reagan did not intend to go through with a last moment repudiation of the treaty that would seriously damage his presidency by provoking Congress, the allies and a large segment of the public.

The Reagan administration will be remembered not so much for its accomplishments in arms control, which were essentially limited to the INF Treaty, but for the remarkable turnabout in the last three years in its attitude toward arms control. This turnabout would not have occurred if the early Reagan administration had succeeded in destroying the existing framework of arms control. Gerard Smith deserves special credit as one of a relatively small group whose persistence and determination blocked this course of events, thus making possible genuine progress when Mikhail Gorbachev introduced Glasnost and peristroika to Soviet arms control policy.

Over the years, Gerard Smith produced a constant flow of "Smithograms," which came as formal memos or scrawled notes, asking questions or suggesting actions on every conceivable facet of our work. Sometimes they sought information, sometimes they sought to stimulate the recipient's lagging imagination. The Smithograms came in a flood when I worked for him at ACDA. When they resumed again at ACA, I knew that Gerry had lost neither his interest nor creativity. One can even expect probing questions about tomes which he had somehow found time to read in his busy life before you discovered the books existed. His Smithograms reveal that, while Gerry is a hard-headed realist with few illusions, he strongly believes that there are solutions to our problems and man should have the wit to find them. He is always searching for new ideas and new solutions to old problems.

In trying to capture the spirit of Gerard Smith, I recall an incident in late 1972, after the ABM Treaty had been ratified, when he called me into his office and confided to me that he had decided to resign his position as Director of ACDA as a result of his intense differences with Henry Kissinger. Genuinely surprised, I protested that he shouldn't do that lest the Agency and the arms control effort collapse without him. He laughed and said that wouldn't happen since ACDA and the people like myself were indispensable to the government, but that he couldn't stay any longer because "his mother had not raised him to be a bureaucrat." Gerry was later outraged when the indispensable bureaucrats who stayed behind were all purged.

Above all, Gerry is not a bureaucrat. He is a man of profound dedication, fierce independence, inventive imagination, and impeccable integrity. He is an aristocrat in the best sense of the word. Unfortunately, there are not many of his kind left; but then, there were never many of his kind to begin with!

Spurgeon M. Keeny, Jr.
Washington, D.C.
January 1989

A Man for All Seasons

It is recorded of Sir Thomas More, saint and statesman, whose felicitous blend of religious faith and practical wisdom illumined an age, that he once said to his son-in-law, William Roper, of his friendship with Henry VIII: "I believe he doth as singularly favor me as any subject in the realm. Howbeit, son Roper . . . if my head could win him a castle in France, it should not fail to go."

More was quite at home in two worlds, practical in political analysis, and principled in his behavior; and when his head finally went, it was not for a castle in France and it was not unforeseen.

Gerard Smith has lived his life in a world of practical deliberations and decisions, but he is, like More, a man of faith as well as a man of reason. It has been given to few men to deliberate significantly on such matters as the survival of the civilized world, and such an awesome task is immeasurably aided by a value system that suggests what a civilized world is. To his work as Director of The Arms Control and Disarmament Agency and Chief of the SALT I delegation particularly he has brought the Christian humanism and the principled intelligence of a Thomas More. His life-long Catholic faith, quietly and firmly practiced, has given him a spiritual depth and a wisdom that admirably complements a natural brilliance.

As a friend of thirty years, I have admired his depth and essential serenity as he has represented our nation at the highest levels. The French describe such a man as *bien dans sa peau,* a man comfortable with himself who knows himself and what he can do. I think Gerard Smith is such a man.

There is, at least, an inadvertent impertinence in presuming to analyze something as personal as one's religious faith. I should prefer to confine myself rather to its principled manifestation in public and private life.

In a privately printed book written a half century ago, our honoree describes with marvelous understatement "the facts of our recent discomfiture" in the Bahamas—the discomfiture being caused by being adrift with six companions in a sometimes turbulent sea in a crippled boat, and coming face to face with the prospect of death. Even as a young man he demonstrated the equanimity of the man of faith, and the ingenuity of the man of reason, quite properly including human ingenuity as a legitimate component of Divine Providence. In the foreword of that little book, *All At Sea,* he quoted a sentence of Horace: *Forsan et haec olim meminisse juvabit,* feeling that pleasurable recollection in future years would come not so much from the fact of having cheated death, but from the fact of emerging from the encounter a stronger man with a greater clarity of vision, and a deeper sense of the reality of life and death.

In the trying hours when rescue was uncertain, he recalls the hymn to Mary, Star of the Sea, sung in chapel at Canterbury school in student days, and the Retreat Master's dictum, oft repeated, that from prayerful silence emerges a different and better person.

A later encounter in the life of Gerard Smith that was to be a significant influence was his meeting the late Father James Keller of Maryknoll, founder of The Christophers, a relatively unstructured organization whose main belief was and is that one person can make a difference in our world by bringing solid values into home and marketplace. His patient skill has indeed made a difference; in the beginning of his book, *Doubletalk,* he quotes St. Francis de Sales:

> A cup of science.
> A barrel of wisdom.
> A sea of patience.

Although the cerebral Gerard Smith is not given to passing enthusiasms, he is receptive to innovative approaches as his diplomatic skills attest. From Canterbury he went to Yale where four years of Latin gave him a happy familiarity with the classical tradition. I think this characterizes the man rather well. He balances a principle of innovation which recognizes what is promising in the future with a principle of tradition which recognizes what is enduring in the past.

Among the services rendered by Gerard Smith to the Church in America was membership on the Board of Trustees of The Catholic University of America. It was there that he met many of the hierarchy to whom he was to serve as consultant in the preparation of the drafts of the Pastoral Letter of the National Conference of Catholic Bishops on War and Peace: *The Challenge of Peace. God's Promises and Our Response.*

As Chief of the United States Delegation to the SALT I negotiations for President Nixon, and Special Presidential Representative for Nonproliferation for President Carter, he was able to provide the authors of this document with invaluable background material for their deliberations. Besides his formal meetings with the committee, he met on several occasions with the Reverend J. Bryan Hehir, Director of the Office of International Justice and Peace of the United States Conference of Catholic Bishops, an important member of the NCCB Committee on War and Peace.

The pastoral letter was inspired by the words of concern expressed in the Second Vatican Council: "The entire human race faces a moment of supreme crisis in its advance toward maturity." The letter attempted from the resources of faith and reason to explore the possibility of a world freed of the threat of nuclear annihilation, by an examination of the long and complex history of the Catholic moral tradition. It did not pretend to be a definitive solution to the problem, but it represented the conviction that the problem was too great to be ignored. The authors felt that ". . . there is no satisfactory answer to the human problem of the nuclear age which fails to consider the moral and religious dimensions of the problems we face."

It may well be true that integrity is not necessarily an oddity on the national and international scene, but there is certainly ground for suspicion that the sinuosity of

Macchiavelli's *The Prince,* rather than the luminosity of the Sermon on the Mount seems to be more in evidence in our world. A reading of Gerard Smith's *Doubletalk,* the story of the SALT I negotiations, reveals his acute discomfort with the ethical consequentialist's justification of the *ragione di stato* as the sole determinant of national policy.

Ambassador Smith is not one given to levity, but he does have a quiet sense of humor at some of the incongruities of the human condition. He narrates, for example, the story of a flight to Moscow from Helsinki, when the negotiators, having had some beer, put the texts of the ABM Treaty and the Interim Agreement on the table. "Whereupon we ran into turbulent air and I had visions of delivering two beer-soaked parchments to our principals. I do not recall whether we put the beer or the documents in a safe place. Perhaps both."

But on the same flight the negotiators wrote some insightful observations on the importance of their arduous work in the months of preparation. Our ambassador wrote: "A day we should all remember with the greatest satisfaction—it may well be the day on which was started one of mankind's great adventures." The ambassador's words remind one of *The Tempest* of Shakespeare wherein the Bard talks of a sea change, a movement into something rich and strange. One might hope that the present conciliatory climate in superpower negotiations will be seen in future years as the culmination of the work of those who embarked on the great adventure years ago.

No mention of the life and values of Gerard Smith would be complete without mention of her who, until her recent untimely death, shared his life, Bernice

Latrobe Smith. She was there as a loving and supportive presence even during the negotiations of the long Helsinki days when a peaceful future was being painstakingly designed.

When I was privileged to give her eulogy in Saint Matthew's Cathedral in Washington on February 6, 1987, I observed that she was almost by instinct an elegant and discriminating woman, and a loving encouraging presence to her best friend and companion of the years. Although she was a private person, I think she would have relished the idea of a *festschrift* to her husband; but it would tell her little that she did not already know.

I would see Gerard and Bernice from time to time on Sunday mornings at the French Mass at the Church of the Epiphany in Georgetown. We also practiced French on long walks at his home in Easton, although at that time he was somewhat distracted by the more ponderous realities of the Russian tongue which he had recently started to learn if not to master.

Lest one get the impression that a 75th birthday and its attendant honors signify the end of a career, one hastens to note that Gerard Coad Smith is still very much alive, and very much involved with The Consultants International Group, Inc.

Moreover there is little indication that our honoree, (whose attitude to travel might be summarized in Cervantes' observation that the road is better than the inn) has lost the interest in traveling about that world he has been so interested in preserving.

But the man of faith and principled reason has indeed reached an important milestone, and he is entitled to look back on the journey thus far with considerable satisfaction before he resumes the journey. Neither SALT, nor its advocate has lost its flavor. And the man for all seasons is still at his job.

Paul Mohan
Washington, D.C.
October 1988

Appendix

On Gerard Smith

While I shall concentrate on my association with Gerry after I had left the government vineyards and then, somewhat later, when he had left Arms Control and Disarmament Agency (ACDA), a few comments on his role at the State Department in the 1950s days are necessary prologue. Gerry and Phil Farley, two hardened bureaucrats, engaged in an uncharacteristic gamble when they accepted into their small office an authentic nuclear neophyte. My only credential was a paper I had written in 1955 at the National War College on the peaceful uses of atomic energy. This was a halcyon time when one and all were all assured that the promise of the atom was such that electricity from nuclear reactors would be so cheap it would not pay to meter the stuff.

The Gerry Smith I recall from the late 1950s was reserved, never putting on display his extensive knowledge of this new and exotic field, managing the office with a light hand, and always available to get us out of trouble when we got in a bind. I suspected at the time that inside that patient exterior existed a kind of amused tolerance mixed with uneasiness about what Schaetzel and Stan Cleveland would be up to next. As we developed the US-Euratom Joint Program Stan and I had confidence that when we needed high level support from Dulles or with the AEC Gerry would be there to help us. He was.

On to the future. In early 1970 I happened to be in Washington on consultation immediately after the Nixon administration took office. I was a holdover from the previous administration and had been Ambassador to the European Community for four years. Just to look in on a friend I called on Gerry in his ACDA office—indeed the same office that had housed Dean Acheson when he was Secretary. Gerry's first remarks were to ask what he could do to see that I stayed on. He didn't wait for me to answer but made notes on the people whom he would approach. That is only one example of an enduring trait: complete loyalty to his associates and friends.

It has been my good fortune in this post-State Department existence to work with Gerry on a number of projects. Some five years ago, with Gerry as chairman, we launched a nuclear nonproliferation project. This initiative was based on the solid premise that Gerry attached a higher priority to the subject and knew more about it than did the government. Without Gerry the project would never have moved beyond the discussion stage. For a man who hides his passions, his deep and eloquent concern over the prospect that nuclear weapons will spread is palpable. He has invested the project with a fervor, a certitude that passivity cannot be tolerated, and contempt for those who share the comfortable illusion that insofar as nothing has happened so far we can all relax. This is part of his exceptional and informed attention to the ultimate danger facing the world: the use of nuclear weapons.

Gerry has not only led the American group but helped us assemble a European group with Johan Holst as its leader. As co-chairman of an American-European conference in Oslo he set the tone by his insistence that the problem would not go away and new efforts would have to be made to arouse international opinion.

Acquired over a forty-year period, Gerry's accumulation of American and foreign politicians and experts, at all levels, has given us an inexhaustible reservoir of talent to draw on. In 1988, as we move into the second phase of the project, we all feel a new sense of urgency. This derives from the political volatility in South Asia, the thinly veiled Israeli weapons program, and the threatening spread of missiles and chemical weapons. Without Gerry this project would have lacked the credibility it now enjoys.

From the perspective of the Volcker Commission on the public service I am struck by how Gerry is one of a vanishing breed, lawyers of exceptional promise who elected careers in government service—men such as Stimson, McCloy, Lovett and Acheson—over less challenging but more crassly rewarding private practice.

Another trait stands out. Many senior government figures on leaving public service are content to take on the passive role of critics of an often absurd passing scene, observers. In contrast Gerry has inspired and led the private arms control effort. In speeches, testimony and articles he has made effective use of decades of experience and on-the-job training. Especially satisfying is to see this man now expressing in the most forceful terms his informed indignation over arrant policies of public leaders, whereas, not very long ago, he was bound by the restraints of his office. There is, unfortunately, no reason to believe that he will lack future stimuli.

J. Robert Schaetzel
Bethesda, Md.
November 1988

On Gerard Smith

I met Gerard Smith through an ad in *The New York Times.* It was 1966. I was then a young man who had been Managing Editor of *East Europe* magazine for a few years and wanted to become involved in West European affairs. The ad seemed to be looking for just that kind of person and so I answered it. This brought me to the offices of the Haire Publishing Company and thence to Gerry Smith.

By happy coincidence, Gerry wanted to start a magazine that would stress the Atlantic dialogue. He had already decided to hire a European as editor, but by the very nature of the transatlantic connection, he wanted an American as deputy editor. I remembered our first meeting very well, for I was struck by his forcefulness and determination. I also remarked on his courtesy and in the years to come this was a marked characteristic in his dealings with me and the other members of the staff of what came to be known as *Interplay,* The Magazine of International Affairs.

Anthony Hartley, who had been a leading writer for *The Economist,* was to be the editor, but for some months, I was in charge of starting up a magazine. Our first problem came with the title. We first called it *Intercontinental,* but then Gerry thought it might be confused with the Intercontinental Hotel chain, so we scrapped that one. We conferred on typefaces; at first, we chose a very harsh typeface, but then changed it before the first issue. Gerry acted as publisher, though he remained in Washington while the magazine offices were located in New York City.

We had small offices at 200 West 57th Street, and our meetings with the publisher were spartan affairs, with sandwiches at our desks as we tried to see if we could make the magazine self-sustaining. When Tony Hartley arrived, we had further discussions about the thrust of the journal, and it was decided to broaden the scope beyond U.S.-European relations and make it a monthly journal of international affairs. Gerry was interested in every detail of the operation: could we sell enough ads? Probably not. Could we find a broad subscription base? Probably yes. After a faltering start with the wrong circulation consultants, we finally got on track and it appeared we would break even by the end of 1971.

The editorial product seemed to me then—and it does now—first rate. Gerry's international connections resulted in a slew of suggestions of who could write from where. Tony Hartley, who was suggested as editor by Francois Duchene, the head of the London Institute of Strategic Studies, seemed to have a raft of high-flying British journalists who could—and did—produce copy under any circumstances. Our covers were designed by the art director of *Art in America.* It was a classy magazine and published any number of writers, such as Bill Pfaff and Henry Fairlie, who later became regular contributors to publications such as *The New Yorker* and *The New Republic.* With Gerry Smith's encouragement we branched out into coverage of U.S. politics and urban affairs.

Occasionally the editors would go to Washington to a dinner party given by the Smiths that turned into a seminar over coffee. I seem to recall an evening with Marshall Shulman talking about the direction of Soviet foreign policy. I remember thinking at the time that such an evening in a private home seemed hard to imagine in New York. Always Gerry Smith showed constancy in his support of the aims of the magazine, and when mistakes were made, he was careful to point them out while urging us to go forward. We had a lot of fun and what we were doing was important. Unfortunately for us, the election of Richard Nixon in 1968 meant that Gerry Smith would be going into the government to negotiate arms control agreements with the Soviet Union and he would have to relinquish his support of *Interplay*. After he left the magazine was sold to others, and things were never the same. The new owners did not share Gerry's vision. I soon left to become Managing Editor of *Foreign Affairs* and Tony returned to England.

The copies of *Interplay* still read very well indeed, and the place it has left in the history of American journals of opinion is a distinguished one. I had a wonderful time helping to make *Interplay* a reality. So did Tony Hartley. And, I hope, so did Gerry Smith.

James Chace,
New York, N.Y.
December 1988